Biology and Radiobiology of ANUCLEATE SYSTEMS

I.
BACTERIA AND ANIMAL CELLS

Proceedings of an International Symposium organized by the Department of Radiobiology, Centre d'Etude de l'Energie Nucléaire (C.E.N./S.C.K.), Mol (Belgium) under the auspices of the Commission of the European Communities (EURATOM) and of the "Relations Culturelles Internationales" (Brussels), held in Mol, June 21-23, 1971.

Biology and Radiobiology of ANUCLEATE SYSTEMS

I. BACTERIA AND ANIMAL CELLS

edited by

Silvano Bonotto   Roland Goutier
René Kirchmann   Jean-René Maisin

Département de Radiobiologie
Centre d'Etude de l'Energie Nucléaire
Mol, Belgium

Academic Press
New York and London
1972

COPYRIGHT © 1972, BY ACADEMIC PRESS, INC.
ALL RIGHTS RESERVED
NO PART OF THIS BOOK MAY BE REPRODUCED IN ANY FORM,
BY PHOTOSTAT, MICROFILM, RETRIEVAL SYSTEM, OR ANY
OTHER MEANS, WITHOUT WRITTEN PERMISSION FROM
THE PUBLISHERS.

ACADEMIC PRESS, INC.
111 Fifth Avenue, New York, New York 10003

United Kingdom Edition published by
ACADEMIC PRESS, INC. (LONDON) LTD.
24/28 Oval Road, London NW1

LIBRARY OF CONGRESS CATALOG CARD NUMBER: 72-78874

PRINTED IN THE UNITED STATES OF AMERICA

In memory of Professor Joseph Maisin

CONTENTS

PARTICIPANTS ix
PREFACE xv

Opening Lecture: Morphogenesis and Synthesis of
Macromolecules in the Absence of the Nucleus 1
    Jean Brachet

JUNE 21
SESSION 1
Chairman: René Thomas

Production of DNA-Less Bacteria 29
    Yukinori Hirota and Matthieu Ricard

Biology and Radiobiology of Minicells 51
    Howard I. Adler and Alice Hardigree

Modifications of Radiosensitivity in Nucleate
and Anucleate *Amoeba* Fragments 67
    Yvette Škreb, Durda Horvat, and Magda Eger

SESSION 2
Chairman: Zénon M. Bacq

Heterogeneity of Membrane-Bound Polyribosomes
of Mouse Myeloma Cells in Tissue Culture 85
    M. Zauderer and Corrado Baglioni

9 S Haemoglobin Messenger RNA from Reticulocytes and
Its Assay in Living Frog Cells 101
    C. D. Lane, G. Marbaix, and J. B. Gurdon

CONTENTS

SESSION 3
Chairman: Maurice Errera

The Squid Giant Axon: A Simple System for the
Study of Macromolecular Synthesis in Neuronal Fibres 117
  Antonio Giuditta

Anucleate Stentors: Morphogenetic and
Behavioral Capabilities 125
  Vance Tartar

Cytoplasmic Damage Leading to Delay of Oral
Regeneration in *Stentor coeruleus* 145
  Brower R. Burchill

JUNE 22
SESSION 4
Chairman: Lucien Vakaet

The Regulation of Protein Synthesis in
Anucleate Frog Oocytes 165
  R. E. Ecker

DNA, RNA, and Protein Synthesis in Anucleate
Fragments of Sea Urchin Eggs 181
  Sydney P. Craig

Synthesis of Closed Circular DNA in Fertilized and
Activated Eggs and in Anucleate Fragments of the
Sea Urchin *Arbacia lixula* 207
  Horst Bresch

PARTICIPANTS

Henri Alexandre, Département de Biologie Moléculaire, Laboratoire de Cytologie et Embryologie Moléculaires, Université Libre de Bruxelles, Rhode-St. Genèse, Belgium

Christa Anders, Institut für Entwicklungsphysiologie der Universität zu Köln, Cologne, Germany

Zénon M. Bacq, Laboratoire de Physiopathologie, Université de Liège, Liège, Belgium

Cyrille Baes, Département de Radiobiologie, C.E.N./S.C.K., Mol, Belgium

Wilfried Baeyens, Département de Radiobiologie, C.E.N./S.C.K., Mol, Belgium

Corrado Baglioni, Massachusetts Institute of Technology, Cambridge, Massachusetts

Danielle Bailly, Département de Radiobiologie, C.E.N./S.C.K., Mol, Belgium

Hubert Balluet, Département de Radiobiologie, C.E.N./S.C.K., Mol, Belgium

Lucille Baugnet-Mahieu, Département de Radiobiologie, C.E.N./S.C.K., Mol, Belgium

Sigrid Berger, Max-Planck-Institut für Zellbiologie, Wilhelmshaven, Germany

Monique Boloukhère, Département de Biologie Moléculaire, Laboratoire de Cytologie et Embryologie Moléculaires, Université Libre de Bruxelles, Rhode-St. Genèse, Belgium

Eliane Bonnijns-Van Gelder, Département de Radiobiologie, C.E.N./S.C.K., Mol, Belgium

Silvano Bonotto, Département de Radiobiologie, C.E.N./S.C.K., Mol, Belgium

PARTICIPANTS

Jean Brachet, Département de Biologie Moléculaire, Laboratoire de Cytologie et Embryologie Moléculaires, Université Libre de Bruxelles, Rhode-St. Genèse, Belgium

Ernst P. O. Brändle, Institut für Biologie der Universität Tübingen, Tübingen, Germany

Horst Bresch, Medizinische Hochschule Hannover, Abteilung für experimentelle Pathologie, Theoretische Institute I, Hanover, Germany

Marc Callebaut, Laboratorium voor Anatomie en Embryologie, Rijksuniversitair Centrum, Fakulteit der Wetenschappen, Antwerp, Belgium

Pol Charles, Département de Radiobiologie, C.E.N./S.C.K., Mol, Belgium

Sydney P. Craig, National Institute of Child Health and Human Development, National Institutes of Health, Bethesda, Maryland

Anne Catherine Dazy, Institut de Biologie Cellulaire Végétale, Université de Paris, Paris, France

Alain Declève, Département de Radiobiologie, C.E.N./S.C.K., Mol, Belgium

Ghislain Deknudt, Département de Radiobiologie, C.E.N./S.C.K., Mol, Belgium

Thomas D'Souza, Biology Division, Bhabha Atomic Research Centre, Trombay, Bombay, India

R. E. Ecker, Biological and Medical Research Division, Argonne National Laboratory, Argonne, Illinois

Maurice Errera, Laboratoire de Biophysique et Radiobiologie, Université Libre de Bruxelles, Rhode-St. Genèse, Belgium

Eugène Fagniart, Département de Radiobiologie, C.E.N./S.C.K., Mol, Belgium

Henri Firket, Institut de Pathologie, Université de Liège, Liège, Belgium

Georg Gerber, Département de Radiobiologie, C.E.N./S.C.K., Mol, Belgium

Yves Gérin, Département de Biologie Moléculaire, Laboratoire de Cytologie et Embryologie Moléculaires, Université Libre de Bruxelles, Rhode-St. Genèse, Belgium

Norberto Gilliavod, Département de Radiobiologie, C.E.N./S.C.K., Mol, Belgium

PARTICIPANTS

Antonio Giuditta, International Institute of Genetics and Biophysics, Naples, Italy

André Goffeau, EURATOM, Université de Louvain, Heverlee, Belgium

Roland Goutier, Département de Radiobiologie, C.E.N./S.C.K., Mol, Belgium

Johannes Hackstein, Zoologisches Institut, Lehrstuhl: Experimentelle Morphologie, Universität zu Köln, Cologne, Germany

Francoise Andrée Hanocq, Département de Biologie Moléculaire, Laboratoire de Cytologie et Embryologie Moléculaires, Université Libre de Bruxelles, Rhode-St. Genèse, Belgium

Maurice N. Harford, Département de Radiobiologie, C.E.N./S.C.K., Mol, Belgium

Raoul Huart, Département de Radiobiologie, C.E.N./S.C.K., Mol, Belgium

Paavo Kallio, Department of Botany, University of Turku, Finland

Fernand Kennes, Département de Radiobiologie, C.E.N./S.C.K., Mol, Belgium

Micheline Kirch, Département de Biologie Moléculaire, Laboratoire de Cytologie et Embryologie Moléculaires, Université Libre de Bruxelles, Rhode-St. Genèse, Belgium

René Kirchmann, Département de Radiobiologie, C.E.N./S.C.K., Mol, Belgium

Klaus Kloppstech, Max-Planck-Institut für Zellbiologie, Wilhelmshaven, Germany

Luc Lateur, Département de Biologie Moléculaire, Laboratoire de Cytologie et Embryologie Moléculaires, Université Libre de Bruxelles, Rhode-St. Genèse, Belgium

Lucien Ledoux, Département de Radiobiologie, C.E.N./S.C.K., Mol, Belgium

Rachel M. Leech, Department of Biology, University of York, Heslington, England

Mark Lemarcq, Département de Biologie Moléculaire, Laboratoire de Cytologie et Embryologie Moléculaires, Université Libre de Bruxelles, Rhode-St. Genèse, Belgium

PARTICIPANTS

Alain Léonard, Département de Radiobiologie, C.E.N./S.C.K., Mol, Belgium

Gerno Linden, Département de Radiobiologie, C.E.N./S.C.K., Mol, Belgium

Paul Lurquin, Département de Radiobiologie, C.E.N./S.C.K., Mol, Belgium

Jean-René Maisin, Département de Radiobiologie, C.E.N./S.C.K., Mol, Belgium

Paul Manil, Faculté des Sciences Agronomiques, Gembloux, Belgium

G. Marbaix, Department of Zoology, University of Oxford, England

Max Mergeay, Département de Radiobiologie, C.E.N./S.C.K., Mol, Belgium

Catherine Micholet-Côté, Département de Radiobiologie, C.E.N./S.C.K., Mol, Belgium

Thomas Palayoor, Cancer Research Institute, Parel, Bombay, India

Evelyne Paulet, Université Libre de Bruxelles, Rhode-St. Genèse, Belgium

Simone Puiseux-Dao, Institut de Biologie Cellulaire Végétale, Université de Paris, Paris, France

Jacques Remy, Département de Radiobiologie, C.E.N./S.C.K., Mol, Belgium

Matthieu Ricard, Institut Pasteur, Paris, France

Gerhard Richter, Institut für Botanik der Technischen Universität Hannover, Hanover, Germany

André Sassen, Département de Radiobiologie, C.E.N./S.C.K., Mol, Belgium

J. F. Scaife, Biology Division, EURATOM, Ispra, Italy

G. G. Selman, Institute of Animal Genetics, University of Edinburgh, Scotland

Akihiro Shima, Laboratory of Radiation Biology, Faculty of Science, University of Tokyo, Japan

Yvette Škreb, Institute for Medical Research, Yugoslav Academy of Sciences and Arts, Zagreb, Yugoslavia

Julia Swinnen-Vranckx, Département de Radiobiologie, C.E.N./S.C.K., Mol, Belgium

PARTICIPANTS

Christian Thomas, Département de Biologie Moléculaire, Laboratoire de Cytologie et Embryologie Moléculaires, Université Libre de Bruxelles, Rhode-St. Genèse, Belgium

René Thomas, Département de Biologie Moléculaire, Laboratoire de Génétique, Université Libre de Bruxelles, Rhode-St. Genèse, Belgium

Monique Tourte, Cytologie et Cytophysiologie de la Photosynthése, Gif-sur-Yvette, France

Yves Tourte, Laboratoire de Biologie Végétale V, Paris, France

Lucien Vakaet, Laboratorium voor Anatomie en Embryologie, Rijksuniversitair Centrum, Fakulteit der Wetenschappen, Antwerp, Belgium

Thérèse Vanden Driessche, Département de Biologie Moléculaire, Laboratoire de Cytologie et Embryologie Moléculaires, Université Libre de Bruxelles, Rhode-St. Genèse, Belgium

Colette van der Ben, Institut Royal de Sciences Naturelles de Belgique, Brussels, Belgium

Dick van der Ben, Institut Royal de Sciences Naturelles de Belgique, Brussels, Belgium

Oscar Van der Borght, Département de Radiobiologie, C.E.N./S.C.K., Mol, Belgium

Constantin Vilău, Oncologic Institute, Bucharest, Rumania

Harry Waris, University of Helsinki, Helsinki, Finland

Claude Watters, Département de Radiobiologie, C.E.N./S.C.K., Mol, Belgium

C. L. F. Woodcock, Harvard University, The Biological Laboratories, Cambridge, Massachusetts

Yoshio Yoshida, Department of Biology, Faculty of Science, Niigata University, Niigata, Japan

Klaus Zetsche, Institut für Biologie der Universität Tübingen, Tübingen, Germany

PREFACE

During the First International Symposium on *Acetabularia,* held in Brussels and in Mol, June 1969, it became apparent that a comparison of the properties of anucleate *Acetabularia* with those of other anucleate cells would not only be very interesting but very useful for a better understanding of many problems concerning the nucleocytoplasmic relationships and cellular differentiation in normal and irradiated cells. Therefore, we decided to bring together, for the first time, scientists working on anucleate systems obtained from bacteria and animal or plant cells.

A three-day symposium on Biology and Radiobiology of Anucleate Systems was organized in the Department of Radiobiology of the Centre d'Etude de l'Energie Nucléaire (C.E.N./S.C.K.) at Mol, Belgium, June 21-23, 1971, thanks to the generous sponsorship of the Commission of the European Communities (EURATOM), the Relations Culturelles Internationales (Brussels), and the C.E.N./S.C.K. The opening addresses were delivered by Mr. M. Mees, representing Mr. J. Goens, General Director of the C.E.N./S.C.K. We thank Professors Z. M. Bacq, J. Brachet, H. Chantrenne, M. Chevremont, M. Errera, H. Firket, P. Manil, and R. Thomas, who contributed, as Scientific Advisers, to the success of the symposium. We are particularly indebted to Mr. E. Brons, Public Relations Officer, and to all the other members of C.E.N./S.C.K. who helped in some way in the organization of the symposium.

The papers presented at the symposium generally concern anucleate systems; a few papers, however, deal with some very real problems (function of membrane-bound polyribosomes, behavior of isolated cellular organelles) of interest to the investigator of anucleate systems.

We hope that these proceedings, published in two volumes, will contribute in some way to a better knowledge of the normal and irradiated cell and of the subtle relations between its nucleus and cytoplasm. Volume I is comprised of an opening lecture delivered by Professor J. Brachet and eleven papers on bacteria and animal cells. Volume II is comprised of sixteen papers relating to plant cells.

Silvano Bonotto
Roland Goutier
René Kirchmann
Jean-René Maisin

Biology and Radiobiology of
ANUCLEATE SYSTEMS
I.
BACTERIA AND ANIMAL CELLS

MORPHOGENESIS AND SYNTHESIS OF MACROMOLECULES IN THE ABSENCE OF THE NUCLEUS[x]

Jean Brachet

Laboratory of Molecular Cytology and Embryology
Department of Molecular Biology
University of Brussels (Belgium)

and

Laboratory of Molecular Embryology
Consiglio nazionale delle Ricerche (Italy)

Introduction

The synthesis of macromolecules such as DNA, RNA and proteins is the result of complex nucleocytoplasmic interactions, which can be studied in a number of ways : biochemical analysis of isolated cell organelles (nuclei, mitochondria, microsomes, etc...), autoradiography (which can be performed on intact cells), comparison of nucleate and anucleate fragments of unicellular organisms or eggs. All these approaches have their advantages and their drawbacks. Biochemical studies afford the best opportunities for a detailed analysis of synthetic processes ; but one is never sure, when one is working with isolated nuclei, for instance, that certain molecules have not leaked out of the nuclei or that cytoplasmic components have not been adsorbed unspecifically on their surface : cytoplasmic contamination of isolated nuclei is often difficult to rule out, while the true intracellular localization of enzymes found

---

[x] Opening lecture of the Symposium

in the " cell sap" often remains open to question. Autoradiography, which can now be used at the electron microscope level without difficulty, gives exceedingly valuable information about the intracellular sites of macromolecular synthesis; but it cannot be easily used for the study of the size of the precursor pool (free amino acids, nucleotides, etc...), so that the specific activity of the neosynthesized macromolecules usually remains unknown. Comparative study of nucleate and anucleate fragments of cells, which is the topic of the present Symposium, would be ideal if the amount of material, which can be obtained by microsurgery, was not often so small that delicate and time-consuming ultra-micromethods have to be used for biochemical analysis. Exceptions to this statement are the alga <u>Acetabularia</u> because of its large size, and the sea-urchin eggs, which can be cut into two halves by centrifugation : this is why we know much more about the biochemistry of anucleate <u>Acetabularia</u> and sea-urchin eggs than about that of <u>Stentor</u> or <u>Micrasterias</u>.

It is clear that, if we wish to understand better nucleocytoplasmic interactions in morphogenesis and synthesis of macromolecules, all available methods should be used : this is why, in the present introduction, the work done on anucleate systems will be integrated in the broader frame of the nucleocytoplasmic interactions in morphogenesis.

Before going into the subject, it is worth pointing out that the three above-mentioned approaches (biochemical analysis of isolated cell organelles, autoradiography and studies on anucleate systems) have led to the same general conclusions : the main function of the nucleus is not, as was believed in the past, energy production or protein synthesis; it is the synthesis of nucleic acids, i.e. DNA and all kinds of RNA's.

In the following, we shall first discuss our present ideas on nucleocytoplasmic interactions in cell differentiation (in embryos, particularly), then examine some of the facts on which these concepts have been based.

ANUCLEATE SYSTEMS: BACTERIA AND ANIMAL CELLS

Theories of differentiation

Many classical experiments have clearly established the importance of <u>cytoplasmic organization</u> during the very early stages of embryonic development : for instance, removal of the "polar lobe" at the trefoil stage of cleavage results, in molluscs, in various defects in the larva; mild centrifugation of a fertilized amphibian egg induces marked microcephaly, while inversion of its polarity, by keeping the egg upside down produces siamese twins. The dorsoventral organization of the amphibian egg can be established at will by the experimenter by simply forcing the egg to rotate in a given plane. As a result of this rotation, the dorsal side becomes visible by the appearance of the so called <u>grey crescent</u>. If the cortex (a few μm thick) of this dorsal side is removed, the egg will cleave and form a blastula, but it will not develop further. If, on the contrary, the dorsal cortex of a fertilized egg is grafted on the ventral side of another fertilized egg, a double embryo will form (Curtis, 1960). In all these experiments, nothing is apparently done to the egg nucleus, while minor alterations of the cytoplasm have far-reaching consequences for further development.

On the other hand, many experiments in which the nucleus has been intentionally injured (in order to produce mutations, aneuploidy, haploidy, hybridization, etc...) definitely show that nuclear integrity is required in order to obtain full and normal development. This effect of the nucleus is hardly conspicuous until the initial period of cleavage is over. But morphogenetic movements, neural induction, organogenesis, tissue and cell differentiation all require the presence of normal, genetically active, nuclei.

Morgan (1934), who was not only an outstanding geneticist, but also a very distinguished experimental embryologist, tried, in 1934, to present a coherent theory of nucleocytoplasmic interactions during development. His basic idea was that the cytoplasmic organization of the egg being heterogeneous, the nuclei (which would all contain the same genes) will be distributed, at the end of cleavage, in regions of the blastula which are chemically different from each other. As a result, genes would become "activated" in certain parts of the embryo and not in other areas. Under the impulse of the activated genes, the cytoplasm would become still more different in the various parts of the embryo : this would lead to further gene activation, and so forth, until the cells have differentiated.

This theory perhaps remains the best one we have when we try to explain embryonic differentiation : in its modern form, it will be said that genes were repressed during cleavage and that they undergo selective derepression during development. Differentiation would result from differential gene activation ; this means that the genes which direct the synthesis of a given protein (haemoglobin, for instance) will be activated (or derepressed) in a given area at a given stage of development. But what causes genes to be " switched" on and off, in a very specific way, remains the great mystery of cell differentiation.

Other theories have been proposed, in recent days, in order to explain cell differentiation : Jacob & Monod (1963) tried to extend their model for gene regulation in bacteria to the differentiating cell. A much more elaborate model was proposed, in 1969, by Britten & Davidson (1969) ; in addition to the fact that the complexity of the model makes it difficult to test experimentally, it suffers from the drawback that it does not take into account cytoplasmic heterogeneity. The lack of interest for possible cytoplasmic controls of gene activity is rather surprising, since Davidson, Allfrey & Mirsky (1964) showed that, in the mollusc <u>Ilyanassa</u>, removal of the polar lobe (a purely cytoplasmic area) affects in a negative way RNA synthesis in the "lobeless" embryo : gene activation is still repressed many hours after the surgical removal of the polar lobe.

Scarano and his associates (1967, 1969) have suggested that specific methylation of DNA might have a determining effect on development. Their experimental findings, though interesting, are not yet quite conclusive. But a major objection to any theory based on chemical modifications of the DNA molecules themselves lies in the fact, which will be discussed later, that the nuclei of the adult still contain, in an active form, all the genes needed for complete differentiation (review by Gurdon & Woodland, 1970). Furthermore, there are many well demonstrated cases of cell dedifferentiation, obtained by simple modifications of the "in vitro" culture conditions.

ANUCLEATE SYSTEMS: BACTERIA AND ANIMAL CELLS

A new and tempting theory of cell differentiation is that of selective gene amplification : the genes which code for the synthesis of a tissue specific protein (haemoglobin for instance) would undergo selective replication in the cells which will produce that protein. It is well known that enormous (over 1000 times) gene amplification occurs in the case of the ribosomal RNA's (r-RNA's) genes during the ovogenesis of the amphibians : the result is a tremendous synthesis of ribosomes during this period. It is less known that Ficq & Pavan have shown, in 1957, that individual genes (which are visible under the microscope in the giant chromosomes of the insect Rhynchosciara) undergo selective replication during larval life. Whether selective gene amplification can account for cell differentiation remains to be seen, but the possibility must be considered very seriously.

Other, less popular, theories shift the emphasis from the nuclear genes to the cytoplasm : they derive, in one way or the other, from the once fashionable plasmagene hypothesis. The main assumption is the presence, in the cytoplasm, of self-reproducible, specific particles (the plasmagenes). An egg would contain a complex population of plasmagenes and differentiation would be the result of a competition between them in the various embryonic areas.

There is no doubt that many of the plasmagenes which were studied some 25 years ago are nothing but viruses. We are now left with very few cytoplasmic organelles which can self-reproduce : the centrosomes of the dividing cell, the kinetosomes of the cilia and flagella, the mitochondria, and the chloroplasts of green plant cells. Since both mitochondria and chloroplasts contain DNA and can be the site of cytoplasmic, non-mendelian mutations, they might theoretically play a role in cell differentiation. In certain invertebrate eggs, of the so called "mosaic" type, unequal distribution of mitochondria is a well-established fact (Reverberi, 1961) : for instance, in Tunicates, there seems to exist a close correlation between an accumulation of mitochondria in certain parts of the egg and the differentiation of the muscles in the tadpole larva. Although this correlation still holds true in centrifuged eggs, it would be unrealistic to explain the facts on the basis of a plasmagene theory. It seems more likely and simpler to assume that muscle differentiation, in an ascidian tadpole, requires a large amount of energy, which would be provided by the mitochondria.

Recently, the plasmagene theory has been reviewed by Bell (1969) and by Curtis (1967). Bell claims the existence in the cytoplasm of "in vitro" cultivated embryonic cells, of DNA-containing particles which he has called, I-somes (I standing for information). The I-somes would be made of "informational" DNA and proteins ; they would, like the plasmagenes, carry genetic information and be capable of self-replication in the cytoplasm. But the reality of the I-somes is very much in doubt now : it is probable that they are just artefacts originating from broken or degenerating nuclei.

The work of Curtis (1965) is of an entirely different kind, since it is based on genetic and not on biochemical considerations. Curtis (1965) observed that, after light injury of the grey crescent in the toad Xenopus, some of the eggs never gastrulate and die as late blastulae. But others develop perfectly and become adults which yield eggs and sperm. They can thus be crossed with normal toads. A cross between a male originating from an egg whose grey crescent (dorsal cortex) had been injured and a normal female gives normal descendance. But the reciprocal cross (between a female originating from an egg whose grey crescent had been injured and a normal male) gives a high proportion of lethals : they die, as if their own grey crescent had been removed, at the late blastula or early gastrula stage. This means that the dorsal cortex apparently has its own heredity : such a "cortical heredity" implies that it must contain self replicating cytoplasmic particles comparable to the plasmagenes. This hypothesis is, in fact, an extension to the fertilized amphibian egg of the ideas of Sonneborn (recent review in 1970), who has produced very strong evidence for the existence of a cortical heredity in Paramecium.

ANUCLEATE SYSTEMS: BACTERIA AND ANIMAL CELLS

A few experimental facts

1. Morphogenesis and synthesis of macromolecules in the total absence of the nucleus

   a. The alga Acetabularia

   (see Brachet & Bonotto, 1970 ; Puiseux-Dao, 1970, for recent reviews)

   The important work of Hämmerling (1934) and his school has demonstrated that, when this giant unicellular alga is cut into two parts, both the nucleate and anucleate halves are capable of extensive regeneration. The "caps" which are formed by the anucleate fragments show all the characteristics of the species to which the nucleus belonged. Further experiments, in which the nucleate and anucleate halves of two different species have been combined, have clearly demonstrated that, in these algae, morphogenesis is controlled by morphogenetic substances of nuclear origin. These morphogenetic substances are distributed along an apico-basal gradient : curiously enough, the basal part of the cytoplasm, which is closest to the nucleus, has the smallest concentration in morphogenetic substances ; the mechanisms which operate for the establishment of the morphogenetic gradient, in Acetabularia, must be more complicated than simple diffusion (Crick, 1970). The morphogenetic substances should be regarded as stable (for at least 2 - 3 weeks) species-specific gene products.
   In biochemical terms, it is likely that the morphogenetic substances correspond to a family of stable messenger RNA's (m-RNA's), capable of directing the synthesis of a large number of proteins, including many enzymes, in the absence of the nucleus. The fact is that the protein content of the anucleate fragments increases 2 to 3 times during the 2 weeks which follow the removal of the nucleus. Experiments in which nucleate and anucleate fragments have been treated with UV, ribonuclease or actinomycin provide strong indirect evidence for the identity of the morphogenetic substances with stable m-RNA's. Furthermore, it can be shown by autoradiography that RNA's synthesized in the nucleus quickly move towards the apex of the alga (where the cap will be formed) ; certain of these RNA species are

so stable that they can still be detected 3 months after a short pulse with a radioactive precursor. Finally, it has been shown that the "template" activity for protein synthesis of the cytoplasmic RNA fraction decreases when the caps are formed. All these findings strongly point out that the morphogenetic substances are indeed stable m-RNA's; one cannot exclude, however, the possibility that proteins synthesized under the control of these stable m-RNA's are the actual agents for morphogenesis, since cap formation is very sensitive to the drugs (puromycin, cycloheximide) which inhibit cytoplasmic protein synthesis.

However, the situation, in <u>Acetabularia</u>, is complicated by the fact that its chloroplasts display a large degree of autonomy towards the nucleus : these chloroplasts contain DNA, various species of RNA's and many proteins. All of them can be synthesized in the absence of the nucleus. Inhibitors of chloroplastic RNA (rifampicin) and protein (chloramphenicol) synthesis sooner or later arrest regeneration in both kinds of fragments. The independence of the chloroplasts towards the nucleus is large, but not complete: chloroplasts increase in number in anucleate fragments, but at a slower rate than in nucleate ones. For this reason, net DNA synthesis stops earlier in anucleate than in nucleate halves, and the same is true for net RNA and protein synthesis. It has also been demonstrated that, in grafts between anucleate and nucleate fragments of two different species, the isozyme pattern of malic dehydrogenase (a chloroplastic enzyme) changes and becomes of the nuclear type.

The role of the mitochondria and mitochondrial DNA in the regeneration of anucleate fragments of <u>Acetabularia</u> might be more important than one thought : ethidium bromide which binds preferentially to circular DNA, inhibits regeneration and abolishes the mitochondrial DNA peak in both nucleate and anucleate halves. The effect of ethidium bromide, like that of cycloheximide and puromycin, is irreversible in anucleate halves, reversible in nucleate ones.

Finally, it has been observed that the regulation of phosphatase activity (or synthesis) is different in the anucleate and nucleate halves : while the latter respond, in the expected way, to culture in a phosphate deficient medium the former are hardly affected by changes in the phosphate content of the culture medium ; regulation mechanisms which operate in nucleate fragments are absent in the anucleate ones.

It should be added that <u>Acetabularia</u> is a convenient material for the study of cytoplasmic effects on the nucleus : if the algae are grown in the dark, in order to stop photosynthesis, the nuclei shrink and the nucleoli lose some of their RNA. In an alga which has just formed its cap, the nucleus divides in order to give rise to many daughter nuclei : but such nuclear division no longer occurs if a large part of the cytoplasm is removed by surgical operation. If, on the contrary, the stalk of an alga with a cap rudiment is grafted on the nucleate half of a very young alga, premature nuclear division can be induced.

In summary, the <u>Acetabularia</u> experiments tell us a number of important facts : morphogenesis can be controlled by m-RNA's which remain active in the cytoplasm for a considerable time ; the chloroplasts display considerable, though incomplete autonomy towards the nucleus ; the latter can be deeply modified (morphology, RNA synthesis, entry in mitosis) by factors which originate from the cytoplasm.

b. Protozoa

Although much interesting work has been done on the regeneration of anucleate Protozoa (see the articles by V. Tartar and by B.R. Burchill in the present Symposium), only <u>Amoeba proteus</u> has been studied with biochemical methods (Brachet & al., 1955).

When amoeba of this species are cut into two with a micro-needle, the anucleate fragments round up and become incapable of feeding and moving ; the nucleate fragments, on the other hand, display normal locomotion and feeding ability. If the two kinds of fragments are kept fasting, both survive during a couple of weeks.

Nucleate and anucleate fragments do not differ much in respect to energy production (respiration, ATP content). However, the anucleate halves are unable to utilize their glycogen and fat reserves, in contrast to their nucleate counterparts. As a result, the protein content of the anucleate halves drops faster, during prolonged fasting, than that of the nucleate fragments. The total RNA content drops in a dramatic way a few days after removal of the nucleus : obviously, RNA synthesis lies under a much closer nuclear control than protein synthesis.

Whether anucleate fragments of <u>A.proteus</u> are capable of independent DNA, RNA and protein synthesis is unfortunately unknown, because all available stocks are contaminated with endosymbionts. Anucleate fragments indeed incorporate thymidine, uridine and amino acids into the corresponding macromolecules ; but it is impossible to decide whether the nucleic acids and proteins synthesized in the absence of the nucleus belong to the amoeba itself or to the intracellular parasites.

Despite this serious drawback, which strongly diminishes the interest of <u>A.proteus</u> as a material for the study of macromolecule synthesis in anucleate cytoplasm, this amoeba has proven extremely useful for the study of nucleocytoplasmic interactions, as shown by the work of Prescott and Goldstein (1969) ; they have injected the nuclei of labeled amoeba into normal, "cold" amoeba and followed the migration of macromolecules (RNA, proteins) from the nucleus to the cytoplasm and vice-versa. The experiments have shown that radioactive nuclear RNA quickly moves to the cytoplasm only a small amount of this cytoplasmic RNA (probably of low molecular weight) of nuclear origin can be re-incorporated in the nucleus. This back and forth movement of the macromolecules is much more important in the case of the proteins: Prescott and Goldstein (1969) were able to demonstrate the existence of so-called cytonucleoproteins, which continuously move from the cytoplasm to the nucleus and back to the cytoplasm. Kinetic studies have shown that one should distinguish between fast and slow moving cytonucleoproteins. As we shall see later, such cytonucleoproteins also exist in eggs and it is likely that they are present in all cells. The significance of such proteins which are synthesized in the cytoplasm and migrate into the nuclei remains unknown: they could play an important role in the control of gene activity, but this remains a speculation for the time being.

c. <u>Eggs</u>

Both sea-urchin and amphibian eggs can undergo repeated cleavage in the absence of the nucleus. In the sea-urchin egg, anucleate fragments can easily be obtained by centrifugation in a density gradient (Harvey, 1936). If these fragments are treated with parthenogenetic agents, asters are formed. In the best cases, after a few replication cycles of the centrosomes, abnormal morulae can be produced. But development never progresses further than abnormal fragmentation of the anucleate halves : there is no hatching and cilia never form. The development of anucleate fragments of sea-urchin eggs is obviously very inferior to the extensive regeneration which can be obtained in Acetabularia after removal of the nucleus. Development is no better in anucleate amphibian eggs, in which both pronuclei have been destroyed by irradiation of the sperm and pricking of the egg nucleus: irregular blastulae can form, but their cells are unable to differentiate further, even if they are grafted in a normal embryo (Briggs & King, 1951). The same kind of result is obtained after heavy irradiation of both gametes : the result is an aneuploid condition, in which the various blastomeres contain nuclei with widely different numbers of chromosomes. Irregular morulae or blastulae are obtained, which die during cleavage.

These experiments clearly show that the cytoplasm of sea-urchin and amphibian eggs contains information sufficient for a few, rather abnormal cleavages ; the presence of a normal nucleus is absolutely required for further development.

Is an anucleate fragment of an egg capable of protein and nucleic acid synthesis ? The answer is "yes" in the case of protein and RNA synthesis, probably "no" in that of DNA synthesis.

Unfertilized sea-urchin eggs synthesize very little proteins, but fertilization induces a fast and considerable increase in protein synthesis (Monroy, 1965). The same sudden increase is observed when anucleate fragments of sea-urchin eggs undergo parthenogenetic activation (Brachet & al., 1963). The anucleate polar lobes of Ilyanassa eggs, isolated at the "trefoil stage" are able to incorporate amino acids into their proteins for at least 24 h (Clement & Tyler, 1967). In the frog Rana pipiens also, enucleate oocytes are able to synthesize proteins when maturation has been induced by hormones (Ecker & Smith, 1968).

All these facts can be explained if we assume that the m-RNA's synthesized during oogenesis are stocked in stable form in the cytoplasm ; under the influence of parhtenogenetic or hormonal stimuli, these m-RNA's would bind to pre-existing ribosomes and form functional polysomes (Burny & al., 1965).

Recent work has shown that various kinds of RNA's can be synthesized by anucleate fragments of sea-urchin eggs , provided that they have been activated by a parthenogenetic agent (Chamberlain, 1970 ; Craig, 1970 ; Selvig & al., 1970). Unpublished experiments carried out in our laboratory on the isolated polar lobe of <u>Ilyanassa</u> tend to support this conclusion. Future work is required to establish whether cytoplasmic RNA synthesis, in such cases, is entirely directed by mitochondrial DNA or not.

There is very little evidence, so far, that DNA synthesis is possible in the absence of the nucleus. One would indeed expect replication of mitochondria to occur in anucleate fragments of eggs, at least when special conditions required for protein synthesis, such as parthenogenetic activation, are satisfied. But convincing experimental proof of this contention is still missing, as will be shown by H.Bresch in this Symposium.

d. <u>Reticulocytes</u>

Enucleation is a natural process during maturation of the red blood cells : the latter no longer synthesize proteins ; but the reticulocytes, which have lost their nucleus and retained their polysomes, are highly specialized in the synthesis of hemoglobin. In the course of their maturation, the reticulocytes progressively loose their polysomes, then their ribosomes and differentiate into adult erythrocytes.

The case of the reticulocytes demonstrates once more that intensive protein synthesis is possible in the absence of the nucleus ; it further shows that this synthesis can be highly specialized so that only one or very few (carbonic anhydrase is also synthesized by the reticulocytes) proteins are made.

The information for the synthesis of the globin chains must be present in the mRNA molecules which link together the ribosomes into polysomes ; this mRNA has been synthesized by the nucleus at an earlier stage and it remains intact, i a stable form (as in <u>Acetabularia</u> or sea-urchin eggs) during many hours. Globin mRNA has been isolated, as a 9 S component from polysomes extracted from reticulocytes (Marbaix & Burny, 1964 ; Chantrenne, Burny & Marbaix, 1967).

It has been demonstrated that it can direct the "in vitro" synthesis of globin. It has recently been shown (Lebleu & al., 1971) that the hemoglobin mRNA is associated with a protein which is required for its binding to the native 40 S ribosomal sub-units. The origin (nuclear or cytoplasmic) of this protein remains unknown.

2. Nucleocytoplasmic interactions during oogenesis and maturation

We have recently reviewed in detail the large and interesting literature which deals with nucleic acid and protein synthesis during oogenesis (Brachet & Malpoix, 1971). All we can do here is to briefly summarize the main facts, limiting ourselves to oogenesis in the toad Xenopus laevis.

DNA replication occurs at the very beginning of oogenesis, at the pachytene stage of meiosis. Two distinct DNA's are synthesized : chromosomal DNA and nucleolar organizer DNA. The latter has a higher density and a base composition which resembles that of the ribosomal RNA's ; for this reason, it is usually called ribosomal DNA (r-DNA). In Xenopus the r-DNA is accumulated, in the nucleus, in a "cap" which is distinct from the chromosomes. It undergoes repeated replication, so that many copies of the nucleolar organizers (which are formed of repeated sequences of the genes coding for the 18 S and 28 S r-RNA's separated by "spacers" which are very rich in guanine and cytosine) are made and that r-DNA is considerably amplified : the amount of r-DNA present in the nucleus of a Xenopus oocyte is about 2000 times higher than in somatic cells of the same species. After replication of the r-DNA, the cap disintegrates and about 1500 nucleoli are formed. The result of this tremendous amplification of the nucleolar organizer genes is the synthesis, by the oocyte, of a huge amount of ribosomes.

However, synthesis of ribosomes does not take place immediately after the r-RNA genes have been amplified in the cap : small Xenopus oocytes (100 to 200 μm diameter) already contain all the copies of the r-RNA genes, but they do not synthesize the 18 S and 28 S r-RNA's. They only synthesize small molecular weight RNA's, with sedimentation constants of 4-5 S . The r-DNA, during this period of growth which precedes vitellogenesis, is clearly in a completely repressed state. Whether the repressor originates from the cytoplasm or not is not known at the present time.

This period of previtellogenesis is followed by vitellogenesis : the nucleoli appear in the nucleus and the ribosomes in the cytoplasm. But the most conspicuous change is the accumulation of yolk platelets in the oocyte : we are dealing here with a cytoplasmic event, since we know that the phosphoproteins which form the bulk of the yolk platelets are synthesized in the liver of the females and carried by the blood to the oocytes where they are absorbed during vitellogenesis. The yolk platelets are more numerous and larger at the vegetal than the animal pole : they are clearly distributed along a vegetal-animal gradient, while the ribosomes and polysomes are distributed along an opposite animal-vegetal polarity gradient (which corresponds to the future cephalo-caudal organization of the embryo and the adult). The establishment of such gradients is of tremendous importance for the future of the embryo, which would have neither head, nor tail if they did not exist. Cytoplasmic factors certainly play a role in their establishment, since there is a larger intake of phosphoproteins at the vegetal than at the animal pole ; but differences in the cell membrane at the two poles of the oocyte, allowing a stronger uptake of the blood phosphoproteins at the vegetal one, might ultimately be under genetic control.

During vitellogenesis, the chromosomes take the well-known lampbrush structure : thousands of loops appear outside the chromosomal axis and actively synthesize DNA like RNA's. Part of these RNA's will cross the nuclear membrane and bind with cytoplasmic ribosomes in order to form polysomes.

But the production of such messenger RNA's (m-RNA's) is quantitatively insignificant (about 3 %) as compared to the synthesis of ribosomes : in fact, the great majority (about 90 %) of the latter is in the form of monosomes (80 S ribosomes) which are not engaged in protein synthesis.

At the end of oogenesis, RNA and protein synthesis progressively decreases. Little is known about the mechanisms, nuclear and cytoplasmic, which control this repression except in the case of r-RNA synthesis : Crippa (1970) recently succeeded in isolating an acidic protein which binds selectively with r-DNA and blocks, after micro-injection, r-RNA synthesis in younger oocytes. This repressor is present in the nucleus of the large, mature oocytes ; but it would not be surprising if, like the histones (Kedes, Gross, Cognetti & Hunter, 1969) it was synthesized in the cytoplasm and tranferred therefrom into the nucleus.

An excellent review article by Smith & Ecker (1970) on the maturation of amphibian oocytes has just been published. Among the many interesting facts which are discussed, some have a direct bearing upon our present problem. For instance maturation (breakdown of the oocyte nucleus, elimination in the cytoplasm of the nucleolar organizers, formation of a maturation spindle) can occur (and is even accelerated) in oocytes treated "in vitro" with a mixture of progesterone and actinomycin ; the latter completely blocks RNA synthesis. On the other hand, puromycin or cycloheximide, which inhibit protein synthesis, make maturation impossible. The breakdown of the nuclear membrane therefore does not require RNA synthesis by the nucleus, but cytoplasmic protein synthesis is essential. In fact, Smith & Ecker (1970) were even able to show that some of the chemical and physiological changes (increase in protein synthesis, modification of the cell membrane) which are induced by progesterone can occur even when the oocyte nucleus has been removed by surgical operation.

An important consequence of maturation, which has been known since a long time and is confirmed by Smith & Ecker (1970), is that the mixing of the nuclear sap and the cytoplasm is essential for cleavage to occur. It is possible that the proteins synthesized during maturation play an important role during cleavage : autoradiography has shown that many cytoplasmic proteins move into the nuclei where they are actively concentrated during cleavage (Ecker & Smith, 1971).

Similar results have been obtained by other authors (see the review by Gurdon & Woodland, 1970) who have studied the penetration of labeled proteins, injected in the cytoplasm, into the nucleus of the oocyte and the cleaving egg. These studies have shown that histones are very actively concentrated in the nucleus of the oocyte, when they are injected in the cytoplasm ; other proteins, like bovine serum albumine, also enter the oocyte nucleus, but they are not concentrated to the same extent.

Finally, one should mention the very important work of

Gurdon (see Gurdon & Woodland,1970) showing that, if nuclei taken from adult tissues are injected in the cytoplasm of an unripe oocyte, they remain practically unchanged ; but, if these nuclei are injected into oocytes which have undergone maturation, they swell as the result of an intake of water and cytoplasmic proteins ; a little later, they lose their nucleoli, replicate their DNA and sometimes even form chromosomes. Maturation therefore leads to an activation or synthesis of the enzymes (of DNA-polymerase type) which catalyse DNA replication. It is very likely that cytoplasmic DNA polymerases move into the nuclei during cleavage, since this has been shown to be the case in sea-urchin eggs.

It is probable that oocytes do not synthesize only proteins and RNA, but also DNA during their maturation (unpublished work by G.Huez & F.Zampetti on starfish oocytes): this very late replication of DNA is of small magnitude and might have something to do with the morphological changes undergone by the chromosomes during maturation.

All this recent work on ovogenesis and maturation leads to a very important conclusion : during these early stages of development, exchanges between the nucleus and the cytoplasm occur in both ways. Some of the RNA's synthesized in the nucleus migrate to the cytoplasm ; but many of the proteins built up in the cytoplasm move into the nuclei.

3. Nucleocytoplasmic interactions in the fertilized egg

We have already mentioned the experiments of Curtis showing that the dorsal cortex (grey crescent) of the Xenopus egg controls later morphogenesis (notably the formation of the dorsal lip of the blastoporus and the subsequent neural induction). It therefore becomes tempting to imagine that this cortical material somehow controls the latter derepression of certain genes in the dorsal half of the embryo (Brachet & Hubert-Van Stevens, 1968).

In order to check this hypothesis, labeled uridine has been injected into Xenopus eggs the dorsal or ventral cortex of which had been damaged by pricking. Our still unpublished results confirm the hypothesis to a large extent. Pricking of the dorsal cortex produces a definite fall (and sometimes an almost complete inhibition) in nuclear RNA syn-

thesis in the dorsal lip of the blastoporus ; the inhibition
closely parallels the extent of the morphogenetic activity.
If, as was the case in two-thirds of the experiments, development comes to an end at the late blastula stage, nuclear
labelling with uridine became practically nil. Damage to the
ventral cortex had, in general, a much less marked effect :
however, in one third of the cases, development also stopped
at the blastula stage and again nuclear RNA synthesis was
hardly detectable. In the control embryos, high synthesis
of DNA and RNA was always observed in the induced neuroblast.

This experimental analysis apparently leads to the expected conclusion that lesions of the dorsal cortex bring
about a more or less delayed inhibition in the synthesis of
RNA in the organizer (just as the removal of the polar lobe
in the experiments of Davidson et al. (1964) delays the onset of nuclear RNA synthesis in the lobeless embryos of
Ilyanassa). But cytological study of the pricked eggs revealed an unexpected complication. Whereas we had thought
that experiments like those of Curtis (1960) provided clearcut evidence of cytoplasmic control over development, cortical lesions often affect the nuclei very strongly : mitotic
abnormalities are numerous and varied, largely as the result of multipolar mitoses. Aneuploidy frequently results
in damaged eggs and we know that it is often lethal. Aneuploidy still occurs when the embryos have reached the age
of 2 to 3 days, as shown by cytophotometric measurements
of the DNA content in the different nuclei of the larvae.
Aneuploidy is much more frequent after dorsal cortical lesions than after ventral cortical lesions. Such findings
could of course explain the results of the genetic experiments carried out by Curtis (1965) without calling for a
hypothetical "cortical heredity" : minor alterations of
the chromosomes (deletions, for instance) would not be detected with cytochemical methods ; yet they could lead to
lethality in the offspring.

These experiments clearly show that light injury of
the cytoplasm can lead to marked alterations of the nuclei.
An effect of the cytoplasm on the nuclei can also be observed after light centrifugation of unsegmented fertilized
eggs : it does not only lead to microcephaly, as shown by
Pasteels (1936), but also to cytological changes in the

gastrula. Neither mitotic abnormalities, nor aneuploidy can be detected after light centrifugation. But the cells at the centripetal pole, highly enriched in lipids, glycogen and RNA, have overdeveloped nucleoli.

On the other hand, at the centrifugal pole, the nuclei belonging to the dense vitelline region are rich in DNA and usually lack nucleoli (unpublished observations). Thus, it can be seen that the relative amount of yolk and of light material plays a determining role in the development of the nucleoli and in the balance between the synthesis of RNA and DNA.

The nature of the cytoplasmic factors which control the activity of the nuclei remains unknown : this important question has been discussed recently by Gurdon & Woodland (1970) who have clearly demonstrated that "old" nuclei, coming from adult liver and brain, swell and synthesize DNA when transplanted into the "young" cytoplasm of an anucleate unfertilized egg. They stop making ribosomal RNA's, the synthesis of which begins again at late blastula, when the host nuclei would have begun to synthesize RNA's. Gurdon & Woodland (1970) have presented very strong evidence in favour of the view that this control (which is not species specific) is at least partly mediated by cytoplasmic proteins which migrate into the nuclei. That such a control mechanism also operates in our centrifuged eggs is extremely likely, but remains to be proved.

Coming back to the egg cortex and its role in morphogenesis, it is known since the famous experiments of Driesch that separated blastomeres of sea-urchin eggs, at the 2-cell stage, can give rise to a complete embryo. The biochemical mechanism of such an embryonic regulation remain unknown : but since the separation of the blastomeres is made possible by the removal of the surface coat (intercellular cement) which links the cells together, one can imagine that this cortical material plays an important regulatory role in the early stages of morphogenesis. In order to test this possibility, we have recently studied the problem from a different angle. Sea-urchin eggs have been disaggregated, during cleavage, in sea water lacking Ca^{2+} and Mg^{2+}; the medium was collected after 2 to 3 h, filtered on Millipore filters and the ionic concentration was brought back to normal. This "conditioned" medium (which contains the surface coat material) was added to normal fertilized eggs : the development stops, either during early cleavage,or at the blastula stage according to the species of the recipient eggs. The blastulae possess abnormally long cilia and they never undergo gastrulation. The active substances present in the conditioned media are partly thermolabile, since gastrulae can be obtained in the presence of heated medium. The effects of the conditioned medium on the synthesis of macromolecules have not been studied yet ; but it is clear that the formation of the nucleoli is enhanced rather than delayed : large nucleoli, very rich in RNA can be seen in the blocked young blastulae, at a stage where nucleoli are not yet visible in control eggs. Simultaneously, DNA synthesis is reduced, as shown cytochemically by the Feulgen reaction. Active contitioned media can be obtained from unfertilized, fertilized and cleaving eggs, but not from blastulae : this shows that the chemical properties of the surface coat undergo changes during cleavage. It can be hoped that continuation of this work will shed some light on the biochemical and morphogenetic properties of the cell surface material.

4. Nucleocytoplasmic interactions during later development.

There is little to add to what has already been said about cleavage : we know that it can occur, to a limited extent, in anucleate or strongly aneuploid eggs. It does not require the synthesis of new RNA's : eggs injected with actinomycin or α-amanitin develop into blastulae ; but the fact that they never go further suggests that the m-RNA's synthesized during cleavage are required for the following stage of development, gastrulation. Protein synthesis on the other hand, is an absolute prerequisite for continuous cell division in cleaving eggs : in the presence of inhibitors, one mitotic cycle is possible at best.

At later stages (gastrulation, neural induction), the nuclei become more and more important : at the beginning of gastrulation, certain genes would be derepressed on the dorsal side and a second gradient, dorso-ventral this time, would be superimposed on the initial gradient of polarity. As a result of the formation of the dorsal-ventral gradient of RNA synthesis, protein synthesis would be more intense in the dorsal than in the ventral regions. These initial gradients, following the cell movements at gastrulation and neurulation, would give birth to the classical cephalocaudal and dorsoventral gradients discovered by the experimental embryologists.

This schematic view, which is in agreement with many experimental observations, places the emphasis of localized gene activation (as in Morgan's theory) at the time of gastrulation. But we do not know what part the cytoplasm plays in controlling gene activation. The work of Tiedemann (1968) suggests that its role might be important in the induction of the nervous system and the mesodermal organs during neurulation : neural induction is accompanied by a derepression of the nuclei, since new species of RNA's are synthesized under the influence of the inducing stimulus. But the latter is very probably mediated by cytoplasmic proteins which move from cell to cell.

Experiments with actinomycin have clearly shown that ectodermal cells are unable to react to the inducing stimulus if they cannot synthesize new RNA species ; but the organizer itself (i.e. the inducing tissue) is hardly affected by actinomycin : it probably already contains the active inducing proteins in the cytoplasm of its cells.

5. Lethal hybrids and mutants

Many hybrid combinations between sea-urchins and amphibians are lethal in early stages of development (advanced blastula or early gastrula). It is difficult to compare the various combinations, because the importance of paternal chromatin elimination and aneuploidy varies widely from case to case. We shall only discuss a few cases, which have been studied with modern methods. First of all, it can be stated that the block in development is not due to the arrest of DNA synthesis : the latter continues, but at a reduced rate, in the blocked hybrids. The nature of the nucleic acids which are synthesized in the _Paracentrotus_ ♀ x _Arbacia_ ♂ hybrid has been studied by Denis & Brachet (1969 a,b; 1970), who used molecular hybridization techniques. When development stops, the hybrid contains three times more maternal than paternal DNA, probably as a result of discrete chromosomal elimination, leading to aneuploidy. But the same hybrids contain three times more _Arbacia_ paternal DNA-

like RNA than maternal Paracentrotus DNA-like RNA. These
findings lead to the conclusion that the transcription of
abnormal paternal DNA is almost 10 times stronger than that
of maternal DNA : the control mechanisms which normally re-
strict DNA transcription are deficient when a foreign   nucleus is introduced in the egg cytoplasm.

Since autoradiography studies have shown that the movement of radioactive RNA from the nucleus to the cytoplasm
is slowed down in the hybrids, it is likely that a control
mechanism operates at the level of the nuclear membrane :
the paternal type of RNA's might, partly or completely, be
retained in the nuclei of the hybrids. Such a retention is
not complete in the Paracentrotus ♀ x Arbacia ♂ combination,
since paternal antigens can be detected in the hybrid blastulae. Retention of RNA in the nuclei might be more effici-
ent in hybrids between American sea urchins, where no paternal antigens can be found ; in these combinations, the
"hatching enzyme" is of strictly maternal type.

A rather different and interesting approach to the
study of the synthesis of RNA in hybrids has been used by
Woodland & Gurdon (1969). They injected a diploid Discoglossus nucleus into an anucleate unfertilized egg of Xenopus
and achieved an arrest in development at the blastula stage.
Normal DNA and RNA synthesis occured up till the time of the
arrest. But no synthesis of r-RNA could be detected : this
shows that the nucleolar organizers of Discoglossus failed
to function in the cytoplasm of Xenopus ; the repression of
r-RNA synthesis, which is a characteristic of cleavage,does
not display species-specificity.

It is clear that the lethal hybrids deserve a much
more complete analysis of the mechanisms which regulate
the synthesis of the macromolecules, at both the transcrip-
tional and post-transcriptional levels. But, besides the
lethal hybrids, some mutations which precociously block development are exceedingly interesting. The best known is
the "nucleolusless" mutation of Xenopus, which has been
studied by Gurdon & Brown (1965) from the biochemical view-
point : these mutants do not form nucleoli, nor r-RNA. Yet
lethality only becomes effective at a relatively late stage,
which shows that the unfertilized eggs of Xenopus have ela-
borated enough ribosomes during oogenesis to cope with the
essential phases of organogenesis without further formation

of ribosomes. The (O) mutant of the axolotl should also be mentioned. Normally, this mutant (when homozygous) is lethal at gastrulation ; but the initial phase can be surmounted if the nuclear sap of oocytes from normal females is injected into the cleaving lethal mutants (Briggs & Cassens, 1966). The latter, as we have recently observed (unpublished), are characterized, like the lethal hybrids, by an abnormal accumulation of RNA in the nuclei during cleavage ; later, DNA synthesis is put to a stop. It would be very interesting to identify the factors which are present in the germinal vesicle and which can correct the lethal effects of the mutation.

Final remarks

This brief introduction to the Symposium gives some indications about the advantages and the limitations of anucleate systems : there is no doubt that the cytoplasm always remains "alive" for a longer or shorter period of time (more than 3 months in Acetabularia) after removal of the nucleus. However, it has lost the main attribute of life , the capacity of reproduction, which can only be restored by the introduction of a nucleus.

Anucleate cytoplasm retains, to a highly variable extent, the properties of the whole cell : in contrast to the amazing capacity of regeneration displayed by the anucleate fragments of Acetabularia, enucleate sea urchin or amphibian eggs can at best, after parthenogenetic activation, undergo irregular and abortive cleavage. In all cases studied so far, protein synthesis continues in the anucleate cytoplasm and there are indications that, in the sea urchin, activated anucleate fragments and normally fertilized eggs produce the same kinds of proteins. But this synthesis of proteins is a consequence of the presence and maintenance , in the anucleate cytoplasm, of m-RNA molecules, of ribosomes and of t-RNA's which had been synthesized by the nucleus before it has been removed. The anucleate cytoplasm will,of necessity, progressively wear out : the preformed m-RNA's , despite their stability, will slowly be degraded and both protein synthesis and morphogenetic potentialities will decrease in intensity. Even in Acetabularia, regeneration of an anucleate half becomes impossible if it has been kept in the dark for 2 or 3 weeks. Furthermore, an anucleate fragment of an egg cannot "invent" new informations : it can synthesize proteins, but only "early" ones, which are normally produced after fertilization or during cleavage. It cannot produce the "late" proteins characteristics of differentiated cells, as was recently shown by O'Dell (unpublished) for the larval and adult antigens of the sea urchin. The capacities of anucleate systems for morphogenesis will therefore depend upon the variety and the stability of the informational molecules they have received from the nucleus.

On the other hand, anucleate systems provide unique opportunities for the study of the replication and synthetic abilities of such semi-autonomous cell organelles as mitochondria, chloroplasts and centrioles. All of them retain some activities (transcription, translation, sometimes even replication) in the absence of the nucleus ; these activities are however lower than in corresponding nucleate system. The comparison between the two systems (nucleate and anucleate) should prove extremely rewarding for our understanding of the cooperation between the nucleus and the mitochondria or the chloroplasts.

Finally, it should be pointed out that the present evidence for the existence of stable m-RNA molecules in anucleate systems is only circumstancial so far : this is a simple, logical hypothesis, but it might not correspond to the whole truth. As pointed out by Bonotto et al. (1971), other alternatives should not be overlooked. It might be well worth looking for the presence, in Acetabularia particularly, of unorthodox enzymes involved in nucleic acid synthesis : RNA replicase and RNA primed DNA polymerase, for instance. If such studies gave positive results, the present interpretation of the facts might undergo deep revision.

References

Bell, E., Nature, Lond., 224, 326-328 (1969).

Bonotto, S., Puiseux-Dao, S., Kirchmann, R. & Brachet, J., C.R. Acad. Sci. Paris, 272, 392-395 (1971).

Brachet, J., Biochim. Biophys. Acta, 18, 247-268 (1955).

Brachet, J., Ficq, A. & Tencer, R., Exptl. Cell Res., 32, 168-170 (1963).

Brachet, J. & Bonotto, S., Biology of Acetabularia, Acad. Press, New York (1970).

Brachet, J. & Malpoix, P., Adv. Morph., 9, 263-316 (1971).

Briggs, R. & King, T., Biological specificity and growth, Princeton University Press, p.207 (1951).

Briggs, R. & Cassens, G., Proc. Natl. Acad. Sci. U.S.A., 55, 1103 (1966).

Britten, R.J. & Davidson, E., Science N.Y., 165, 349-357 (1969).

Burny, A., Marbaix, G., Quertier, J. & Brachet, J., Biochim. Biophys. Acta, 103, 526-528 (1965).

Chamberlain, J.P., Biochim. Biophys. Acta, 213, 183-193 (1970).

Chantrenne, H., Burny, A. & Marbaix, G., Prog. Nucleic Acid Res. Mol. Biol., 7, 173-194 (1967).

Clement, M.C. & Tyler, A., Science N.Y., 158, 1457-1458 (1967)

Craig, S.P., J. Molec. Biol., 47, 615-618 (1970).

Crick, F., Nature Lond., 225, 420-422 (1970).

Crippa, M., Nature Lond., 227, 1138-1140 (1970).

Curtis, A.S.G., J. Embryol. exp. Morph., 8, 163-174 (1960).

Curtis, A.S.G., Arch. Biol., 76, 523-546 (1965).

Curtis, A.S.G., The cell surface. Its molecular role in morphogenesis, Academic Press (1967).

Davidson, E.H., Allfrey, V.G. & Mirsky, A.E., Proc. Natl. Acad. Sci. U.S.A., 52, 501-508 (1964).

Denis,H. & Brachet,J., Proc.Natl.Acad.Sci.U.S.A., 62, 196-201 (1969 a).

Denis,H. & Brachet,J., Proc.Natl.Acad.Sci.U.S.A., 62, 438-445 (1969 b).

Denis,H., Europ.J.Biochem., 13, 86-93 (1970).

Ecker,R.E. & Smith,L.D., Dev.Biol., 18, 232-249 (1968).

Ecker,R.E. & Smith,L.D., Dev.Biol., 24, 559-576 (1971).

Ficq,A. & Pavan,C., Nature Lond., 180, 983-984 (1967).

Gurdon,J.B. & Brown,D.D., J.Molec.Biol., 12, 27-35 (1965).

Gurdon,J.B. & Woodland,H.R., Curr.Topics Dev.Biol., 5, 39-70 (1970).

Hämmerling,J., Arch.Entw.Mech.Org., 131, 1 (1934).

Harvey,E.B., Biol.Bull.Mar.Biol.Lab.Woods Hole, 71, 101 (1936).

Jacob,F. & Monod,J., Cytodifferentiation and macromolecular synthesis, Ed.Locke, p.30, Acad.Press.,New York (1963).

Kedes,L.H., Gross,P.R., Cognetti,G. & Hunter,H.L., J.Molec.Biol., 45, 337-351 (1969).

Lebleu,B., Marbaix,G., Huez,G., Temmerman,J., Burny,A. & Chantrenne,H., Europ.J.Biochem., 19, 264-269 (1971).

Marbaix,G. & Burny,A., Biochem.Biophys.Res.Commun., 16, 522 (1964).

Monroy,A., Chemistry and Physiology of fertilization, Ed. Holt, New York, (1965).

Morgan,T.H., Embryology and Genetics, New York University Press, (1934).

Pasteels,J., Dalcq,A. & Brachet,J., Mém.Musée Royal Histoire Naturelle, Série II, p.822 (1936).

Prescott,D.M. & Goldstein,L., Ann.Embryol.Morphog.Suppl.1, 181-188 (1969).

Puiseux-Dao,S., Acetabularia and cell biology, London Logos Press (1970).

Reverberi,G., *Adv.Morph.*, 1, 55-101 (1961)

Scarano,E., Iaccarino,M., Grippo,P. & Parisi,E., *Proc.Natl. Acad.Sci.U.S.A.*, 57, 1394-1400 (1967).

Scarano,G., *Ann.Embryol.Morph.* Suppl.1, 51-61 (1969).

Selvig,S.E., Gross,P.R. & Hunter,A.L., *Dev.Biol.*, 22, 343-365 (1970)

Smith,L.D. & Ecker,R.E., *Curr.Topics Dev.Biol.*, 5, 1-38 (1970).

Sonneborn,T.M., *Proc.Roy.Soc.B.*, 176, 347-366 (1970).

Tiedemann,H., *J.Cell Physiol.*, 72, Suppl.1, 129-144 (1968).

Woodland,H.K. & Gurdon,J.B., *Dev.Biol.*, 20, 89-101 (1969).

JUNE 21

SESSION 1
Chairman: René Thomas

PRODUCTION OF DNA-LESS BACTERIA

Yukinori HIROTA and Matthieu RICARD

Service de Génétique cellulaire de l'Institut Pasteur et du Collège de France, Institut Pasteur, Paris, France.

Introduction

The cell division cycle of E.coli involves the following events : in a cell, which contains one or two nuclei, the DNA is replicated so that two (or four) nuclei are formed : the cell elongates and the newly formed DNA replicas progressively separate from each other. A septum is produced near the equatorial plane so that two cells, each one containing half the DNA content of the mother, are formed and separate from each other (Robinow, 1960). These cells processes are well adjusted to a wide range of culture conditions (Maaløe and Kjeldgaard, 1966).

Results

Thermosensitive mutants of E.coli K 12 altered in the processes of cell division

For several years, we have been working on the mechanism of regulation of cell division in E.coli, and taking a genetic approach to this problem. In some of the mutants isolated, we have found a production of DNA-less bacteria (Hirota and Jacob,1966). Independently, Adler and his associates have found another type of DNA-less cell, called mini-cells (Adler,Fisher, Cohen and Hardigree, 1967).

A series of thermosensitive mutants of E.coli K 12 has been isolated which grow normally at low (30°C) but not at high (41°C) temperature, and characterization of the mutants has been made. Among the thermosensitive mutants that we have examined, a series of mutants defective in different aspects of cell division were found : such mutants may be defective either in initiation of DNA synthesis (DnaA,C) or in elongation of the DNA chain (DnaB, D) ; separation of sister chromosome or DNA partition (Par A,B) and septum formation (FtsA,B,C,D : Min A, B) (Hirota, Ricard and Shapiro, 1971)(Fig.1). Photographies of mutant cells of each class stained by the Piéchaud's technique (1954) are shown in Plate 1. Among the mutants defective in cell division, several classes were found to segregate cells having no DNA under the non-permissive growth conditions. The production of DNA-less bacteria from a series of thermosensitive mutants has already been reported (Hirota and Jacob, 1966 ; Hirota, Ryter and Jacob, 1968 a ; Hirota, Jacob, Ryter, Buttin and Nakai, 1968 b).

Thermosensitive mutants altered in the different processes of cell division have been mapped, and the preliminary results are shown schematically in Fig.2.

Process of abnormal cell division and production of DNA-less bacteria.

Several different classes of thermosensitive mutations producing DNA-less bacteria have been found, and these are summarized in the following sections :

1. Cell division accompanying defective DNA synthesis : defect in elongation of the DNA chain (DnaB, D) or in initiation of DNA synthesis (DnaA, C).

The physiology and genetics of some of these thermosensitive mutations have been studied and have already been described (Hirota, Ryter and Jacob, 1968 ; Hirota, Mordoh and Jacob, 1970 ; Mordoh, Hirota and Jacob, 1970). We will not discuss these here.

Under non-permissive growth conditions, DNA synthesis in these mutants is inhibited, either immediately or after the completion of DNA cycle (DnaA, C) (DnaB, D) : these cells still divide once or twice for 1 h without segregating DNA-less cells. After this 1 h lag, filament form and the segregation of DNA-less cells eventually begins.

This lag appears to be independent of any residual DNA

synthesis, i.e. DNA-chain elongation mutants which stop DNA synthesis immediately after the temperature shift also show an hour lag. The number of nuclei per cell is reduced during the time of cell division : thereafter the septation of cells is inhibited, and filamentous cells are formed. In the filamentous cells, the nuclear mass remains at the central region leaving both cell-termini without DNA, as is also the case when DNA synthesis is stopped by thymine starvation (Cohen and Barner, 1954). In contrast to thymine starvation, during which cell division is completely stopped, many thermosensitive mutants of DNA synthesis continue residual septations for several generations at the reduced rate. This abnormal septation occurs, not between the DNA copies as in the wild type parent, but at the two extremities of a filament on either side of a nuclear mass. Thus three daughter cells can be formed from a filamentous mother cell. The middle one of the three daughters contains all the DNA of the mother, and the other two have none. The number of DNA-less cells in the population increases linearly with time, and reaches 30 to 50 % of the population after several hours. Septation takes place at sites about $1 \sim 2$ μ from the cell extremities and the size of DNA-less cells is roughly the same as that of a normal cell. The size of the nucleated filaments is variable (Plate 1) (Fig.1 and 3).

Furthermore, it has been shown that the two poles of a filament produce DNA-less cells, and one pole can produce more than one DNA-less cell. Therefore, a single filament must be capable of producing several septa during its residual growth (Hirota et al., 1968 a).

2. Cell division accompanying defective DNA partition (Par A, B).

As in the mutant of DNA synthesis, many mutants of this class produce DNA-less bacteria. In this case at the non-permissive temperature, the separation of sister chromosomes is inhibited although the DNA synthesis continues, and filamentous cells are formed after one or two generations. Many mutants of this class continue cell division, and produce DNA-less bacteria of about normal size. Some characteristics of this class of mutants were described previously (Hirota et al.,1968 a ; Hirota, Ricard and Shapiro, 1971)(Fig.1 and 3).

Mutations required in the production of DNA-less bacteria

As we have reported previously (Hirota et al., 1968a, 1968b), two events are required for the production of DNA-less bacteria. One is the formation of filaments with extremities free of nuclei (obtained for instance by the mutation listed above), the other is the Div mutation which causes the filaments to divide thereafter.

1. The expression of the thermosensitive defect

The cessation of either DNA synthesis or partition of sister DNA's at the non-permissive temperature, is somehow involved in the production of DNA-less bacteria. In a collection of ten independent thermoresistant revertants isolated from a thermosensitive mutant which produces DNA-less cells at 41°C, all showed less than few percents DNA-less cells in the culture under the same conditions.

In general terms, one could say, therefore, that any block in DNA synthesis or DNA partition which produces filamentous cells having nuclear mass(es) remaining at the center could be a cause of the production of DNA-less bacteria.

2. The role of Div-mutations

There are reasons to assume that a secondary mutation is required for the production of DNA-less bacteria.
- Although many thermosensitive mutants segregating DNA-less bacteria have been identified, a large number of other mutants at the same locus do not segregate DNA-less cells. From those mutants, it is possible to isolate secondary mutants which now segregate DNA-less bacteria (plate 2) (Hirota et al., 1968a).
- Such secondary mutants have some selective advantages over the original mutant : the secondary mutants survive better under the various culture conditions including the non-permissive temperature.

The secondary mutations do not manifest themselves when alone, but their existence in some thermoresistant strains can be demonstrated . For instance, if a thermosensitive mutation (DnaB of HfrT42) is introduced by P1 transduction into two different thermoresistant strains

(PA 505 and PA 505 MB△101 C), all the thermosensitive transductants from PA 505 MB△101 C produce DNA-less cells at 41°C, while those from PA 505 do not. One such Div mutation, DivB, was mapped closely to the ArgH locus. Although we have not yet mapped precisely all the Div mutants, we know that some other Div mutations do not map at the same locus (Hirota et al., 1968 a, 1968 b).

3. DivC mutation

Recently, we found a new Div mutation, called DivC, by isolating mitomycin-resistant strains from thermosensitive mutants which originally do not segregate DNA-less cells.

If the original mutants of DNA synthesis or DNA partition do not produce DNA-less cells at 41°C, it might be due to an inhibition of septation resulting from DNA degradation. Since mitomycin is known to form cross-links on DNA and to inhibit DNA synthesis (Iyer and Szybalski, 1963; Weissbach and Lisio, 1965) one should be able to select a class of Div mutants using mitomycin-resistance. This implies that the effect of the Div mutation involves the DNA repair process which may be coupled with cell division in some way.

Seven different mitomycin-resistant types have been reported (Greenberg, Mandell and Woody, 1961). The parental strains we have used are defective either in DNA synthesis (thermosensitive chain-elongation effect : DnaB-CRT 1425), or DNA partition (thermosensitive DNA strand separation defect : ParB-MFT 100). From these thermosensitive mutants, we have isolated a series of mitomycin-resistant mutants. Among the resistant mutants, from ParB, 4 out of 16 became thermoresistant but the rest remained thermosensitive. 3 of the 12 thermosensitive-mitomycin-resistant mutants were found to segregate DNA-less cells at the non-permissive temperature (Table 1 a).

When the original thermosensitive mutant, MFT 100, is grown at 41°C, cell division is arrested, filaments form, and the colony forming capacity is lost exponentially. However, the mitomycin-resistant derivatives, MT 100-DivC, continue cell division and segregate DNA-less bacteria (Fig.4a,4b). They also survive better at 41°C. The chromosome location of the DivC mutation was mapped near the His loci.

Among eight independent mitomycin-resistant mutants of DnaB, all remaining thermosensitive, three were found to segregate DNA-less cells, while the others and the parental strain do not (Table 1 b).

4. Mini-cell production : mutant classes having abnormal septation sites (Min A,B).

Adler and his associates have reported the production of small spherical cells of about 0.3 - 0.5 µ diameter having no DNA, called mini-cells (Adler et al., 1967). Among the mutants we have isolated, some thermosensitive mutants are found to map at two different loci, called MinA and MinB. One MinA was found to produce mini-cells at both permissive and non-permissive temperatures (Hirota et al., 1968 a ; Hirota et al., 1971). The other MinB produces mini-cells only at the non-permissive temperature and low cell concentration. The mini-cell producers synthesize DNA and the separation of sister DNA's continues normally under the non-permissive growth conditions. Mini-cells are produced by abnormal septation occuring very closely to the poles of the filaments, and in addition, septation may also occurs at the normal site. The abnormal septation causes, therefore, the production of very small cells having no DNA and multinucleated filaments of various sizes (Fig.1 and 3).

5. Min-mutations

Genetics of Min-mutations is still preliminary. A mapping result of Adler's Min locus is already reported (Taylor,1970).

Three mutations controlling the thermosensitivity, the UV sensitivity and the mini-cell production have been separated by cross in the case of our MinA mutant. The preliminary mappings of the mutation, suggest the location closed to lac locus as assigned by Adler, are schematically shown in Fig.2. Some of the thermoresistant recombinants still segregate mini-cells. Their cell-size seems to be longer than that of the wild type. This suggests, therefore, that abnormal septation may disturb the process of normal septation. In the other hand, the TS mutation associated with the MinB mutation has been found to be co-transducible by P 1 with ArgH.

Properties of DNA-less bacteria

Properties of DNA-less cells have been reported (Hirota and Jacob, 1966 ; Hirota et al., 1968 a, 1968 b ; Adler et al., 1967). Comparing the results given by Adler and his associates and by us, the characters of DNA-less bacteria are seen to be remarkably similar, although these cells are segregated by different mechanisms. The character of DNA-less bacteria can be summarized as follows :

1. The first class of DNA-less cells that we have characterized are roughly of the size of a normal cell. They are segregated from the mutants of DNA replication and DNA partition. They are homologous in size while nucleated filaments are not. The other class of DNA-less cells, mini-cells, are spherical in shape and very small in size (Adler et al., 1967 ; Hirota et al., 1968 a). DNA-less cells can be purified by filtration and a population of cells can be obtained which contains less than one percent of nucleated contaminants (plate 2).
2. DNA-less cells do not contain any material stainable by the Piéchaud technique (specific for DNA) or any amount of DNA detectable by chemical methods (Burton procedure). Neither do they contain any radioactivity when they are produced from bacteria grown at 30°C for several generations in the presence of ^{3}H-thymidine, then raised at 41°C in the same medium. Otherwise, by their chemical composition (protein, and RNA) or morphological structures (cell wall, cell membrane and ribosomes), they appear very similar to normal cells (plate 3).
3. DNA-less cells respire and oxidize glucose normally. When DNA-less cells were segregated from bacteria previously induced for β-galactosidase, the DNA-less cells contained the full activity. They contain DNA polymerase I in normal quantity. However, little if any DNA-dependent RNA polymerase was detected in these cells and presumably it segregates with DNA.
4. DNA-less cells do not synthesize DNA, RNA, proteins or enzymes such as β-galactosidase.
5. When infected with T6, DNA-less cells produce very few or no phage particles at 41° or 30°C.

Discussion

There are two interesting aspects of the thermosensitive cell system producing DNA-less bacteria. The first one concerns the regulatory mechanism of cell division. There is a well-known set of mutants which do not form septa either after small doses of radiation (Howard-Flanders et al., 1964 ; Adler and Hardigree, 1964 ; Walker and Pardee, 1967) or temperature shifts (Van de Putte, Van Dillewijn, Rörsch, 1964 ; Kohiyama, Cousin, Ryter and Jacob, 1965). The arrest of DNA synthesis after either thymine-starvation or the addition of DNA inhibitors (Cohen and Barner, 1954 ; Gross, Deitz and Cook, 1964) causes the inhibition of septum formation. Consequently, filamentous cells are formed.

These facts suggest the existence of a genetically controlled sequence of signals triggering the series of reactions which connect the processes of the cell cycle : DNA replication, DNA partition and cell division. Such a model predicts the existence of a series of pleiotropic mutations which, in some way, alter the sequence of signals. One may therefore speculate that those signals involve some specific triggering proteins, which can be altered by a secondary mutation in such a way that a thermosensitive step can be by-passed to a later step. Then the whole cell cycle may be able to reoperate again through the altered circuit which results from the secondary mutation.

DivC and the other mitomycin-resistant mutation which are able to correct the thermosensitive mutations are, therefore, especially interesting.

The second aspect concerns the determination of the septation site. It is remarkable that the DNA-less cells produced by DnaA,B,C,D - Div or ParA,B-Div are of the same size as normal wild type bacteria in contrast to mini-cells (Adler et al., 1967).

It should be noted that both poles of a filament can produce DNA-less cells, and one pole can produce more than one such cell during its residual growth. In this system, the mechanism must therefore be preserved which correlates increase in cell length with septum formation and thereby determines the position of the septum with respect to the bacterial extremities.

In the light of the replicon model of Jacob, Brenner and Cuzin (1963), one may deduce a model for site determination of septa in the system producing DNA-less bacteria (Fig.5).

The properties of DNA-less cells are precisely those predicted for cells with no genetic information : they are unable to synthesize any polymer the synthesis of which is known to be DNA dependent. Such processes are now quite well understood in bacteria, and we will not discuss them further here.

Acknowledgments

We thank Dr. F.Jacob for the illuminating discussions during the course of this work, and Misses C.Barnoux and M.C.Ganier for their excellent technical assistance in some experiments.

We thank Dr.J.Johnston for his help in preparing this manuscript.

This investigation was supported in part by grants from the "Commissariat à l'Energie Atomique", the "Centre National de la Recherche Scientifique", the "Délégation Générale à la Recherche Scientifique et Technique" (n° 6600275) and the "National Institutes of Health" (AI 07885)

References

Adler,H.I. and A.A.Hardigree (1964) Analysis of gene controlling cell division and sensitivity to radiation in Escherichia coli. J.Bacteriol., 87, 720-726.

Adler,H.I., W.D.Fisher, A.Cohen and A.A.Hardigree (1967) Miniature Escherichia coli cells deficient in DNA. Proc.Nat.Acad.Sci.U.S.A., 57, 321-326.

Burton,K. (1956) A study of the conditions and mechanism of the diphenylamine reaction for the colorimetric estimation of deoxyribonucleic acid. Biochem.J.,62,315-323.

Cohen, S.S. and H.D. Barner (1954) Studies on unbalanced growth in E. coli. Proc.Nat.Acad.Sci.U.S.A., 40, 885-893.

Greenberg, J., J.D. Mandell and P.L. Woody (1961) Resistance and cross-resistance of Escherichia coli mutants to antitumour agent mitomycin C. J.Gen.Microbiol., 26, 509-520.

Gross, W.A., W.H. Deitz and T.M. Cook (1964) Mechanism of action of nadilixic acid on Escherichia coli. J.Bact., 88, 1112-1118.

Hirota, Y. and F. Jacob (1966) Production de bactéries sans DNA. C.R.Acad.Sci., 263, 1619-1621.

Hirota, Y., A. Ryter and F. Jacob (1968 a) Thermosensitive mutants of E. coli affected in the processes of DNA synthesis and cellular division. Cold Spring Harbor Symp. Quant.Biol., 33, 677-693.

Hirota, Y., F. Jacob, A. Ryter, G. Buttin and T. Nakai (1968 b) On the process of cellular division in Escherichia coli. I. Asymmetrical cell division and production of deoxyribonucleic acid-less bacteria. J.Mol.Biol., 35, 175-192.

Hirota, Y., J. Mordoh and F. Jacob (1970) On the process of cellular division in Escherichia coli. III. Mutations affecting the initiation of DNA synthesis. J.Mol.Biol. 53, 369-387.

Hirota, Y., M. Ricard and B. Shapiro (1971) The use of thermosensitive mutants of E. coli in the analysis of cell division. J.Cell.Physiol., in press.

Howard-Flanders, P., E. Simon and L. Theriot (1964) A locus that controls filament formation and sensitivity to radiation in Escherichia coli K 12. Genetics, 49, 237-246.

Iyer, V.N. and W. Szybalski (1963) A molecular mechanism of mitomycin action : linking of complementary DNA strands Proc.Nat.Acad.Sci.U.S.A., 50, 355-362.

Iyer, V.N. and W. Szybalski (1964) Mitomycins and porfiromycin : chemical mechanism of activation and cross-linking of DNA. Science, 145, 55-58.

Jacob,F., S.Brenner and F.Cuzin (1963) On the regulation of DNA replication in bacteria. Cold Spring Harbor Symp.Quant.Biol., 28, 329-348.

Kohiyama,M., D.Cousin, A.Ryter and F.Jacob (1966) Mutants thermosensibles d'Escherichia coli K12. I. Isolement et caractérisation rapide. Ann.Inst.Pasteur, 110, 465-486.

Maaløc,O., and N.O.Kjeldgaard (1966) Control of macromolecular synthesis. W.A.Benjamin Inc.,Ed. New York.

Mordoh,J., Y.Hirota and F.Jacob (1970) On the process of cellular division in Escherichia coli. V.Incorporation of deoxytriphosphates by DNA thermosensitive mutants of Escherichia coli. Proc.Nat.Acad.Sci.U.S.A., 67, 773-778.

Piéchaud,M. (1954) La coloration sans hydrolyse du noyau des bactéries. Ann.Inst.Pasteur, 86, 787-793.

Robinow,C.F. (1960) In "The cell" J.Brachet and A.E.Mirsky, Ed., Academic Press, New York, vol. 4, p.45.

Taylor, A.L. (1970) Current linkage map of Escherichia coli. Bacteriol.Rev. 34, 155-175.

Van de Putte,P., J.Van Dillewijn and A.Rörsch (1964) The selection of mutants of Escherichia coli with impaired cell division at elevated temperature. Mut.Res., 1, 121-128.

Walker,J.R. and A.B.Pardee (1967) Conditional mutants involving septum formation in Escherichia coli. J.Bacteriol., 93, 107-114.

Weissbach,A. and A.Lisio (1965) Alkylation of nucleic acids by mitomycin C and porfiromycin. Biochemistry, 4, 196-200.

Table 1 - Phenotypes of mitomycin-resistant strains derived from MFT 100 (ParB) and CTR 1425 (Dna-B)

a. MFT 100

Number of independant mutants	Thermo-sensitivity	Segregation of DNA-less bacteria (%)
3	TS	Yes (15 - 40 %)
4	TR	No ($\sim$ 3 %)
9	TS	No ($\sim$ 3 %)
Parental strain	TS	No ($\sim$ 2 %)

b. CRT 1425

Number of independant mutants	Thermo-sensitivity	Segregation of DNA-less bacteria (%)
5	TS	Yes (10 - 25 %)
3	TS	No ($\leq$ 2 %)
Parental strain	TS	No ($\leq$ 2 %)

Bacterial culture in exponential phase at 30°C in broth was shifted to 41° for 3 h. Then nuclei were stained by the Piéchaud's technique (1954) and 200-400 cells were examined under the microscope to measure the percentage of DNA-less cells among the whole population.

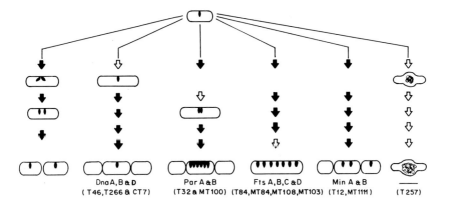

Fig. 1 - A schematic representation of the morphological alteration of thermosensitive mutants altered in the different steps of cellular division.
The open arrow ⇩ signifies the process(es) altered.
The closed arrow ⬇ signifies the normal process (es) of cellular division.
The genes, and the typical mutations are indicated under the each figures, respectively.

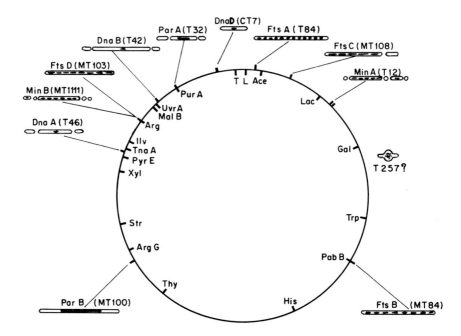

Fig. 2 - A preliminary genetic map of the mutations altered at the different steps of cellular division.

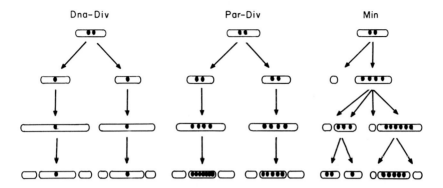

Fig. 3 - A schematic representation of the processes of segregation of DNA-less cells and mini-cells. Segregation of mini-cells is due to an alteration of the min gene; while DNA-less cells are produced, in our case, by the association of a Div mutation with a thermosensitive mutation either in a gene controlling DNA synthesis or in a gene which normally assures the separation of the newly formed DNA-copies.

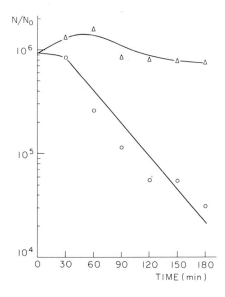

Fig. 4.a. - Survival curve, at 41°C, of the original thermosensitive mutant MFT 100 (defective in DNA partition). o—o, and of the mitomycin-resistant thermosensitive derivative MFT 1002 △ — △ which produces DNA-less cells at high temperature.

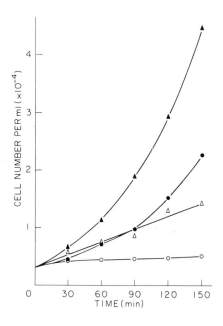

Fig. 4.b. - A cell count (measured with a Coulter counter) of the mutant MFT 100 (at 30°C ●—● and at 40°C o—o), and of its mitomycin-resistant thermosensitive derivative MFT 1002 which produces DNA-less cells at 41°C, △—△, but not at 30°C, ▲—▲.

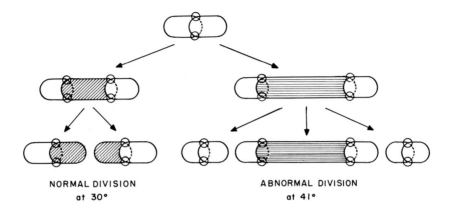

Fig. 5 - A possible model for the mechanism of the segregation of DNA-less cells deduced from the replicon model (Jacob, Brenner and Cuzin, 1963). At 30°C, the cell membrane is normal, functional septation sites are formed, and after they have been separated by the elongation of the cell envelope, they operate to make the cell divide.
At 41°C, the cell membrane is altered and the newly formed septation sites are inactive, therefore filaments form. However, the septation sites formed at 30°C before the temperature shift now located at the cell extremities, are still functional when the Div mutation is present, and they cause the production of DNA-less cells.

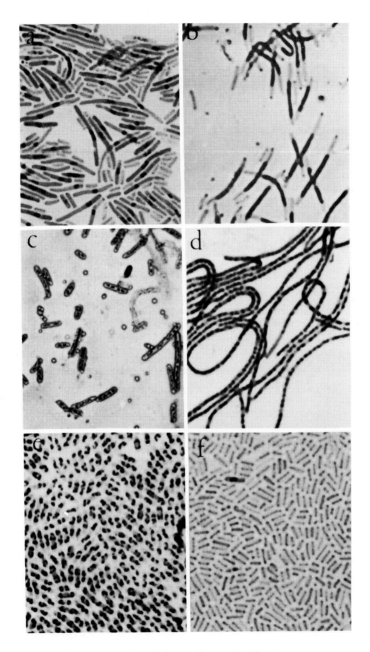

Explanation for Plate 1 on following page.

Plate 1 - Nuclear staining (Piéchaud, 1954) of wild type
and mutant bacterial cells examined under light
microscope.

    a) mutant CRT 46 grown in broth for 4 h at 41°C

    b) mutant PAT 32 grown in broth for 4 h at 41°C

    c) mutant CRT 12 grown in broth for 4 h at 41°C

    d) mutant PAT 84 grown in broth for 2 h at 41°C

    e) wild type CR 34 grown in broth for 4 h at 41°C

    f) purified suspension of non-nucleated cells.
A culture of CRT 46 incubated for 4 h at 41°C
was purified and then stained.

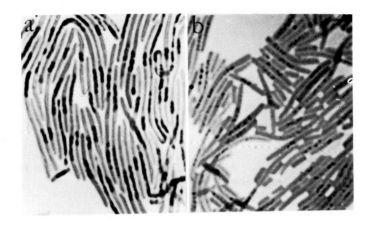

Plate 2 - Nuclear staining (Piéchaud, 1954) of thermosensitive mutants defective in DNA elongation

a. mutant PAT 42 (DnaB) grown in broth for 4 h at 41°C.

b. mutant PAT 421 (DnaB DivB) which is derived from PAT 42 grown in broth for 4 h at 41°C. It is able to segregate DNA-less cells.

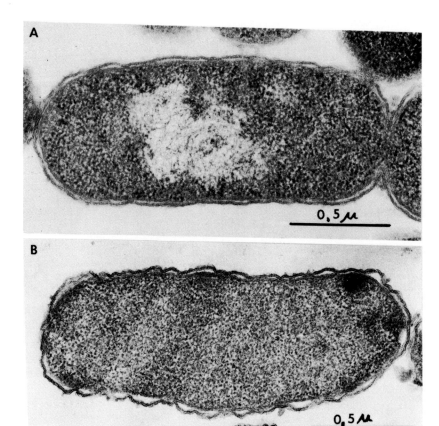

Plate 3 - a. Electron micrograph of a thin section of a nucleated normal E.coli K 12

b. Electron micrograph of a thin section of a non-nucleated bacterium from a purified preparation.

No DNA fibrils can be seen in the DNA-less cell (b), but other morphological features are indistinguishable from those of a normal cell (a). Electron micrograph was taken in the Pasteur Institute by Dr. Antoinette Ryter.

BIOLOGY AND RADIOBIOLOGY OF MINICELLS

Howard I. Adler and Alice A. Hardigree

Biology Division, Oak Ridge National Laboratory[x]
Oak Ridge, Tennessee 37830, U.S.A.

Abstract

Several years ago we isolated a cell division mutant of the bacterium Escherichia coli which has proven to be valuable in several radiobiological and biological studies (Adler et al., 1966, 1967). The striking property of the mutant strain is its ability to produce large numbers of small anucleate structures. These are formed by division-like events occurring very close to the poles of the cylindrical Escherichia coli cells (Fig.1). The "minicells" formed in this process do not have appreciable quantities of DNA but do retain many of the characteristics one would predict for an anucleate but otherwise intact bacterium.

The minicell-producing strain was isolated during a search for mutants of Escherichia coli K-12 with altered radiation sensitivity. We had initiated this search in the course of our studies on the genetic control of radiation response in this organism. The mutant was isolated from Escherichia coli K-12 P678, a widely used recipient strain which had been well characterized by Jacob and Wollman (1961). P678 had been treated with triethylene melamine, a potent mutagen, and among the clones surviving this treatment the minicell-producing strain was found and designated as P678-54. This strain was originally isolated not for the

---

[x] Operated by Union Carbide Corporation for the U.S. Atomic Energy Commission. Paper read by title.

property of producing minicells but because it seemed to have undergone an alteration in radiation response. The observation that it produced minicells was made only after careful microscopic observation of the culture by time-lapse cinematography. An individual cell of Escherichia coli that made minicells had been previously observed by Hoffman and Frank (1963). The unique feature of our isolate is that most of the cells in the culture can produce minicells and the minicells accumulate during the growth of the culture in large quantities. This fact permits their isolation and characterization.

Production and isolation of minicells

Escherichia coli P678-54 produces minicells in all growth media that have been tested. Minicells are produced only during the logarithmic phase of growth, that is, during the period when the cells are actively undergoing division. The minicells accumulate in such a culture, and the final yield of minicells consists of units formed at all stages of the logarithmic phase. By the use of time-lapse cinematography we have attempted to determine if there is any pattern to the way in which cells choose between normal division and minicell-yielding division. It has not been possible to observe any such pattern except to note that at a given point in time a particular cell will either produce a minicell or undergo a normal cell division. The two events have never been observed to occur simultaneously in a single cell.

There are basically two methods that have proven to be effective in the separation of minicells from the parental cells. One of these is based on centrifugation techniques and the other makes use of the resistance of minicells to lysis by penicillin (Levy, 1970). Centrifugation is very effective, particularly if sucrose or glycerol gradients are employed. By the use of such gradients it is possible to obtain minicell preparations of very high purity, less than one contaminating cell per 10^5 minicells. The penicillin lysis method makes use of the fact that minicells are nongrowing structures and resist the lytic action of penicillin. The method offers options to the density gradient centrifugation approach since it does not expose the cells to sucrose or glycerol, which may interfere with some of the cellular processes of interest. By the penicillin method an approximately 500,000 fold purification is obtained. By the use of these techniques it is possible to harvest minicells in milligram or even gram quantities. The medium used for growth before separation seems to be relatively immaterial as far as yield of minicells is concerned. In almost all growth media so far tested, the minicell titer in a stationary phase culture is approximately 1 to 3 times that of the viable cell titer.

Biology of minicells and the minicell-forming strain

Isolated minicells are observed as approximately spherical objects in electron microscope studies and appear to be surrounded by normal wall and membrane. They also seem to contain a cytoplasm of the same density and general characteristics as that observed in normal cells. In the vast majority of minicells, no nuclear material can be observed; however, Tudor, Hashimoto and Conti (1969) have observed nuclear bodies in some minicells obtained from a late log phase. Such DNA-containing minicells have not been observed in stationary phase cultures. It appears likely that a DNA-containing minicell produced during logarithmic phase is capable of sufficient growth so that if allowed to continue incubation it will achieve a size sufficient to separate with the normal cells in differential centrifugation.

Minicells isolated from a sucrose density gradient are capable of respiration for at least 8 hours. During this period they produce acid and gas from sugar substrates (Black, 1967). They do not grow or exhibit any evidence of significant protein, RNA or DNA synthesis. This is understandable in view of the fact that they contain little or no DNA, messenger RNA or certain of the enzymes known to be involved in early steps in protein synthesis. It has been established, however, that extracts of minicells are capable of protein synthesis if an artificial message is added (Fralick et al., 1969).

Our general hypothesis regarding the nature of the mutation in P678-54 is as follows. We believe that a mutation has occurred which interferes with the normal spatial relationship between some essential division triggering signal and the potential division planes of the cell. In normal cells there seems to be a region near the midpoint of the cell which can respond to a division triggering stimulus. In the minicell-forming strain, there may be secondary sites located near both poles of the cell which can also be triggered to form a crosswall and septum. It is possible that these sites exist in normal cells but are somehow blocked from responding to the division initiating stimulus. The genetic control of minicell formation is complex and seems to involve genes in at least two regions of the chromosome. One region is between proC and purE close to the lon locus ; another is the pdx-pyr region (Roozen and Curtiss, personal communication). The genetic evidence is, however, incomplete at this time.

Our reasons for thinking that the minicell-yielding process consists of a misplacement of the division plane depends on the following evidence :

1. We have explored a variety of growth conditions and there is no exception to the fact that minicell formation only occurs under conditions in which normal cell division also occurs. For example, logarithmic phase cultures produce minicells ; stationary phase cultures do not.

2. A double mutant by incorporating the lon^- allele into the minicell-forming strain yields an organism which makes minicells only under conditions where the strain can undergo normal divisions as well. If the double mutant is exposed to radiation, a condition known to prevent further normal cell divisions, minicell formation is also prevented.

3. A double mutant made between the conditional division mutant BUG-6 isolated by Reeve, Groves and Clark (1970) and a minicell-forming strain yields an organism which produces minicells only under conditions that allow division in BUG-6. Under the nonpermissive conditions the double mutant forms filaments and also fails to form minicells.

4. We have made many time-lapse films of the growth and division of minicell-forming cultures and have observed that a given cell at a particular point in time will either be making a minicell or undergoing a normal divi-

sion. The two processes have never been observed to occur simultaneously. If a cell forms a series of minicells from one or both ends, the body of the cell gets unusually long since it has not been undergoing any normal divisions during this period.

For these and other reasons we are operating on the hypothesis that the biochemical events leading to minicell formation are identical or very nearly the same as those leading to a normal cell division. Apparently in a minicell-forming mutant these events can take place at sites other than the midpoint of the cell. It should be noted, however, that not any point on the membrane or wall of the cell can form a crosswall and membrane. The minicells come only from the very ends of the rod-shaped organism and tend to be rather uniform in size, suggesting that the total number of potential division planes in an E.coli cell is limited to three, the usual one near the midpoint of the cell and two others near the poles. These latter sites may in fact be sites intended for cell division in the upcoming generation and in minicell mutants may be triggered prematurely.

Enzyme distribution between cells and minicells

One obvious use for minicells stems from the fact that they represent samples of material from the end regions of growing Escherichia coli cells. By comparing the concentrations of different enzymes and other molecules in minicells and the nucleated portion of the cell, it is possible to develop some concept of the spatial organization of important molecules in this rod-shaped bacterium. Shortly after the isolation of the minicell-forming strain, Hurwitz and Gold (1968) suggested some interesting relationships between the bacterial DNA and the concentration of enzymes involved in DNA metabolism. They were able to establish that the DNA dependent RNA polymerase could not be found in purified minicells but seemed to be retained in the nucleated portion of the cell. The DNA dependent DNA polymerase, on the other hand, was found in minicells in almost normal concentrations. DNA methylase was not found in minicells whereas RNA methylase was found in the normal concentrations. Other enzymes associated with nucleic acid synthesis and repair have also been studied in this way and will be discussed in a later section.

Michaels and Tchen (1968) have investigated the distribution of polyamines between minicells and the nucleated portion of cells. It has been suggested by several investigators that polyamines play a significant role in the control of nucleic acid synthesis and function. Michaels and Tchen were able to establish that there is no difference in the concentration of either putrescine or spermidine in minicells as compared to normal cells. This finding suggests that at least in <u>Escherichia coli</u> there is probably no special intimate relationship between polyamines and the DNA. Other workers have investigated the distribution of enzymes normally found in the periplasmic region of the bacterium. A number of these surface enzymes display higher specific activities in extracts of minicells than in extracts of the nucleated parental cells. These enzymes include alkaline phosphatase, cyclic phosphodiesterase, acid hexosemonophosphatase 5' nucleotidase and ribonuclease I. In contrast a number of internal cytoplasmic enzymes showed elevated or similar specific activities in extracts of nucleated cells versus extracts of minicells (Dvorak et al., 1970).

In addition to enzymes, other macromolecules and cellular structures have been studied using the minicell-forming mutant. For example, two groups of authors have established that the membrane and wall components of growing cells are not made preferentially at the poles of rod shaped <u>E.coli</u> cells since the newly synthesized materials are not found preferentially concentrated in the minicells derived from actively growing cultures (Green and Schaechter, 1971 ; Wilson and Fox, 1971). These observations are strong evidence for the concept that wall and membrane synthesis in <u>Escherichia coli</u> takes place at multiple sites along the length of the growing cell and is not preferentially carried on either at the poles or central region. Minicells have also been used to investigate distribution of membrane sites that can bind to DNA. It has been established that minicells isolated immediately after they have been conjugated with normal DNA-containing donors have DNA bound to their membranes suggesting that potential attachment sites exist near the poles of the bacterium and are incorporated into minicells (Shull et al., 1971).

Distribution of plasmids

Escherichia coli cells may contain, in addition to the chromosome, small DNA structures known as plasmids which contain information that helps determine the phenotype of the cell. Although evidence exists that at least some plasmids segregate with the chromosome in an orderly manner, others may be free in the cytoplasm or linked to the membrane at sites other than that to which the chromosome is presumed to be attached. In the last few years several groups of workers have made use of the minicell-producing strains for the isolation and characterization of plasmids. For example, Kass and Yarmolinsky (1970) established that the sex factor F'gal and F'λ could be isolated from minicells formed during the growth of strains harboring these plasmids. Only a small fraction of the minicells formed contain the F' plasmid; however, these minicells can donate the plasmid when conjugated with appropriate recipients. This observation establishes that, at least under some conditions, the sex factor may be dissociated from the chromosome. Such dissociation does not prevent it from being effectively transferred during conjugation. In similar experiments Inselburg (1970) and Inselburg and Fuke (1970) established that the plasmid-controlling synthesis of colicine E1 can also be segregated into minicells. The ColE1 DNA in this study was found to be predominantly covalently-closed circular molecules which can undergo one or more cycles of replication in the minicells. These workers were able to prepare striking electron micrographs demonstrating the plasmids in the act of replication. Levy and Norman (1970) have established that certain R factors which control resistance to a variety of antibiotics also segregate into minicells. They also established that at least some of the DNA segregated in this manner was in a circular form and could be transferred from the minicells to normal cells as efficiently as in the usual cell to cell matings. More recently Roozen et al., (1971) and Levy (1971) have established that the R factor in a minicell can direct the synthesis of additional DNA, RNA and protein. Cohen and his coworkers (1971) have made use of R factor-containing minicells to establish that at least some of the DNA in minicells is present in the form of catenated molecules.

Work in our own laboratory is extending many of the observations discussed in this section. Particular attention is being directed at the characterization of RNA and protein made in plasmid-containing minicells and the possibility of using minicells for the isolation of plasmids and defined chromosomal segments (Fenwick et al., 1970, 1971 ; Roozen et al., 1970). In this latter connection, it has been demonstrated that populations of minicells containing any one of several plasmids or F's can be prepared and the DNA isolated from them is free from contaminating chromosomal DNA.

Conjugation with minicells

The minicell-producing mutant was isolated from an F^- K 12 strain of Escherichia coli. Examination by electron microscopy of the walls and membranes of minicells suggested that these walls and membranes are identical to those present on the normal DNA-containing cells. Therefore it was predicted that minicells should be capable of forming effective pairs with donor K 12 strains and might be capable of incorporating DNA from such donors. These predictions were tested and reported in a series of papers from our laboratory (Cohen et al., 1967, 1968 a, 1968 b).

It was first established that minicells from an F^- strain could receive radioactively labeled DNA from an F^+ donor. The DNA isolated from minicells after such conjugations was found to be a mixture of single and double stranded material, and it was established that single stranded material could be converted to double stranded structures in the minicell. These observations strongly suggested that during conjugation with an F^+ strain minicells acted as recipients of the F factor itself and that the F factor was transferred as a single stranded molecule which was later converted to a double stranded one. Further experiments using Hfr and F' donors helped to clarify the situation. Transfer from Hfr donors or donors containing long F's resulted in the accumulation of short lengths of single stranded material in minicells with no conversion to double stranded molecules, in contrast to the observation with F^+ donors. The DNA transferred by Hfr or F' donors probably contains only information near the origin of the chromosome which may be insufficient to direct strand doubling in the minicell whereas, in the case of matings with F^-l donors, the entire F molecule is transferred and can direct such synthesis.

Fralick and Fisher (1970) have further characterized the single stranded DNA obtained from minicells after conjugation with Hfr donors by hybridization studies. Their work suggests that the transferred DNA represents an unique strand and supports earlier findings of Rupp and Ihler (1968) and Ohki and Tomizawa (1968).

Furthermore, it suggests that by a judicious selection of donor strains, hybridization techniques and enzymatic treatments, it may be possible to use minicells for the isolation of specific genes (Fralick, 1970 ; Roozen et al., 1970).

Radiobiology of minicells and minicell-forming strains

We have pointed out in an earlier section that the minicell-forming strain was isolated because of its unusual radiation response. Minicell production was an unselected aspect of the mutant's phenotype. We had been carrying out investigations on the genetic control of radiation resistance (Adler and Copeland, 1962). Since many investigators were successfully unraveling the genetic control of sensitivity to ultraviolet light, we chose to look for mutants with an altered response to ionizing radiation ; strain P678-54 has this characteristic. It can be seen from Fig.2 that it is considerably more resistant to ionizing radiation than its parent, but its response to ultraviolet light is not noticeably different. An obvious, but as yet unanswered, question is to establish if the radiation response of P678-54 and the minicell-forming characteristic are due to the same genetic alterations. We have on occasion isolated from P678-54 clones that have lost the ability to form minicells but have retained radiation resistance.

This is not, however, conclusive proof that the two characteristics are determined by different loci and at this point in time we must say that the evidence is incomplete.

With regard to the radiobiology of isolated minicells, we can discuss several interesting observations. Experiments performed by Drs. J.K.Setlow and A.Cohen and reported in Cohen et al., (1968 b) several years ago indicated that the isolated DNA-less minicells had little or no photoreactivating enzyme. These experiments were performed by preparing extracts of the minicells and testing the ability of these extracts to repair ultraviolet damaged transforming DNA. More recently Paterson and Roozen (submitted to J.Bacteriol.) have examined the ability of minicells harboring plasmids to carry out repair of both UV and ionizing radiation damage. They have observed that such plasmid-containing minicells are as capable as nucleated cells in the rejoining of single strand breaks induced in DNA by ionizing radiation. The minicells, however, have a reduced capacity for the photoenzymatic repair of UV-induced pyrimidine dimers and are completely incapable of excising dimers induced in the plasmid DNA by ultraviolet light. These authors conclude that the polar regions of E. coli cells probably contain few or no UV specific endonuclease molecules but may contain some photoreactivating enzyme if a plasmid is in the region. Put another way, they suggest that UV specific endonuclease molecules are confined to the central region of the cell whereas photoreactivating enzyme molecules are associated with DNA and will be found wherever the DNA is found. With regard to the enzymes involved in repair of single strand breaks, it is impossible to say at this stage if these enzymes are found in minicells because they are freely migrating cytoplasmic enzymes or because they are associated with the plasmid DNA. The inability to make this decision stems from the fact that at present there is no obvious way of assaying for the presence of the single strand repair function in minicells not harboring plasmids.

Minicells are being used to investigate at least one other radiobiologically important question. In an earlier section we pointed out that the best available evidence indicates that minicells contain preformed sites in their membrane which can serve as attachment points for newly introduced single stranded DNA. It is conceivable that such attachment sites represent an important target for radiobiological damage. Minicells provide an opportunity for irradiating these sites in the absence of DNA and then assaying to see if their ability to bind DNA has been damaged. Experiments have been performed along these lines (Fralick, 1970). The results indicate that the ability of minicell membrane to bind DNA is an extremely radiation resistant phenomenon. Such findings do not support a hypothesis which suggests that these membrane binding sites are an important radiobiological target. Experiments with a similar purpose have been performed using minicells that have received DNA as a result of conjugation prior to radiation exposure (Shull et al., 1971). In this case the question is asked, "Can irradiation bring about the detachment of DNA from the membrane ?" The answer seems to be that even at very high doses little or no effect is observed. It should be borne in mind that these experiments on the sensitivity of the attachment site depend in large measure on the sensitivity and specificity of the magnesium sarkosyl gradient technique for the determination of membrane attachment of DNA (Tremblay et al., 1969). Since doubt exists about the extent to which this technique critically measures biologically functional membrane-DNA attachment sites, we must hold some reservation as to the meaningfulness of the irradiation experiments just described.

Summary

Minicells of E.coli are a spontaneously formed anucleate sample of an organism that is widely used in a variety of biological investigations. They provide us with an opportunity to investigate the capabilities of this organism when it contains no DNA or small genetically defined DNA structures. To date they have allowed the investigation of several interesting biological and radiobiological questions. We anticipate that they will continue to be useful.

References

Adler, H.I. and Copeland, J.C. (1962) - Genetic analysis of radiation response in Escherichia coli. Genetics, 47, 701-702.

Adler, H.I., Fisher, W.D. and Stapleton, G.E. (1966) - Genetic control of cell division in bacteria. Science, 154, 417.

Adler, H.I., Fisher, W.D., Cohen, A. and Hardigree, A.A. (1967) - Miniature Escherichia coli cells deficient in DNA. Proc.Nat.Acad.Sci.U.S.A., 57, 321-326.

Black, J.W. (1967) - Growth characteristics of miniature Escherichia coli cells deficient in DNA. Master's Thesis, Univ. of Tenn.

Cohen, A., Allison, D.P., Adler, H.I. and Curtiss III, R. (1967) - Genetic transfer to minicells of Escherichia coli K 12. Genetics, 56, 550.

Cohen, A., Fisher, W.D., Curtiss III, R. and Adler, H.I. (1968) DNA isolated from Escherichia coli minicells mated with F^+ cells. Proc.Nat.Acad.Sci.U.S.A., 61(1), 61-68.

Cohen, A., Fisher, W.D., Curtiss III, R. and Adler, H.I. (1968) The properties of DNA transferred to minicells during conjugation. Cold Spring Harbor Symp.Quant.Biol., 33, 635-641.

Cohen, S.N., Silver, R.P., McCoubrey, A.E. and Sharp, P.A. (1971) - Isolation of catenated forms of R factor DNA from minicells. Nature, New Biology, 231, 249-251.

Dvorak, H.F., Wetzel, B.K. and Heppel, L.A. (1970) - Biochemical and cytochemical evidence for the polar concentration of periplasmic enzymes in a minicell strain of Escherichia coli. J.Bacteriol., 104, 543-548.

Fenwick, R.G., Jr., Roozen, K.J. and Curtiss III, R. (1970) - RNA and protein synthesis in plasmid-containing minicells of Escherichia coli K 12. Genetics, 64, s 19.

Fenwick, R.G., Jr., Roozen, K.J. and Curtiss III, R. (1971) - Characterization of RNA transcribed from Col and R plasmids. Bact.Proc., p. 51, 71st Ann.Meeting of ASM, Minneapolis, Minn., May 2-7.

Fralick, J.A., Fisher, W.D. and Adler, H.I. (1969) - Polyuridylic acid-directed phenylalanine incorporation in minicell extracts. J.Bacteriol., 99, 621-622.

Fralick, J.A. (1970) - Radiation, genetic and biochemical studies of the Escherichia coli minicell. Doctoral dissertation, Univ. of Tenn.

Fralick,J.A. and Fisher,W.D. (1970) - Hybridization studies on DNA transferred to minicells. Bact.Proc., 70th Ann. Meeting of ASM, Boston, Mass., April 26 - May 1, 1970.

Green,E.W. and Schaechter,M. (1971) - How is the cell membrane distributed among progeny cells ? Bact.Proc., p.158, 71st Ann.Meeting of ASM, Minneapolis, Minn., May 2 - 7.

Hoffman,H. and Frank,M.E. (1963) - Time lapse photomicrography of the formation of a free spherical granule in an Escherichia coli cell end. J.Bacteriol., 86, 1075 - 1083.

Hurwitz,J. and Gold,M., In Cohen et al., 1968. Cold Spring Harbor Symp.Quant.Biol., 33, 635-641.

Inselburg,J. (1970) - Segregation into and replication of plasmid deoxyribonucleic acid in chromosomeless segregants of Escherichia coli. J.Bacteriol., 102, 642-647.

Inselburg,J. and Fuke,M. (1970) - Replicating DNA : Structure of colicin factor E1., Science, 169, 590-592.

Jacob,F. and Wollman,E.L. (1961) - " Sexuality and the Genetics of Bacteria" Academic Press, New York.

Kass,L.R. and Yarmolinsky,M. (1970) - Segregation of functional sex factor into minicells. Proc.Nat.Acad.Sci.U.S.A., 66, 815-822.

Levy,S.B. (1970) - Resistance of minicells to penicillin lysis : A method of obtaining large quantities of purified minicells. J.Bacteriol., 103, 836-839.

Levy,S.B. (1971) - Studies on R factors segregated into E. coli minicells. Trans. N.Y.Acad.Sci., (in press).

Levy,S.B. and Norman,P. (1970) - Segregation of transferable R factors into Escherichia coli minicells. Nature, 227, 606-607.

Michaels,R. and Tchen,T.T. (1968) - Polyamine control of nucleated and enucleated Escherichia coli cells.J.Bacteriol., 95, 1966-1967.

Ohki,M. and Tomizawa,J. (1968) - Asymetric transfer of DNA strands in bacterial conjugation. Cold Spring Harbor Symp.Quant.Biol , 33, 651-658.

Paterson,M.C. and Roozen,K.J. (1971) - Photoreactivation, excision repair, and strand rejoining in plasmid-containing minicells of Escherichia coli K 12. (submitted to J.Bacteriol.).

Reeve,J.N., Groves,D.J. and Clark,D.J. (1970) - Regulation of cell division in Escherichia coli : Characterization of temperature-sensitive division mutants. J.Bacteriol. 104, 1052-1064.

Roozen,K.J. and Curtiss III,R. (personal communication) Abstr. Ky-Tenn. Branch Meeting, ASM, Asheville, N.C., October 3-4, 1969.

Roozen,K.J., Fenwick,R.G.,Jr., and Curtiss III,R. (1970) - Isolation of plasmids and specific chromosomal segments from Escherichia coli K 12. Informative Molecules in Biological Systems, Ed. L.Ledoux, North Holland, Amsterdam, p. 249-266.

Roozen,K.J., Fenwick,R.G.,Jr. and Curtiss III,R. (1971) - Synthesis of ribonucleic acid and protein in plasmid-containing minicells of Escherichia coli K 12. J.Bacteriol., 107, 21-33.

Rupp,W.D. and Ihler,G. (1968) - Strand selection during bacterial mating. Cold Spring Harbor Symp.Quant.Biol., 33, 647-650.

Shull,F.W., Fralick,J.A., Stratton,L.P. and Fisher,W.D. (1971) - Membrane association of conjugally transferred DNA in Escherichia coli minicells. J.Bacteriol., 106, 626-633.

Tremblay,G.Y., Daniels,M.J. and Schaechter,M. (1969) - Isolation of a cell membrane-DNA-nascent RNA complex from bacteria. J.Mol.Biol., 40, 65-76.

Tudor,J., Hashimoto,T. and Conti,S.F. (1969) - Presence of nuclear bodies in some minicells of Escherichia coli. J.Bacteriol., 98, 298-299.

Wilson,G. and Fox,C.F. (1971) - Membrane assembly in Escherichia coli. II.Segregation of preformed and newly formed membrane proteins into cells and minicells. Biochem.Biophys.Res.Comm., (in press).

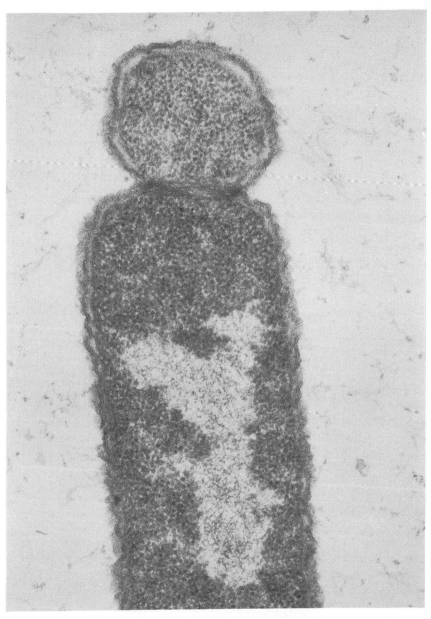

Fig. 1 - Thin section electron micrograph of Escherichia coli K 12 P678-54 demonstrating a minicell yielding division. Photograph courtesy of David Allison. Magnification approximately 76,000.

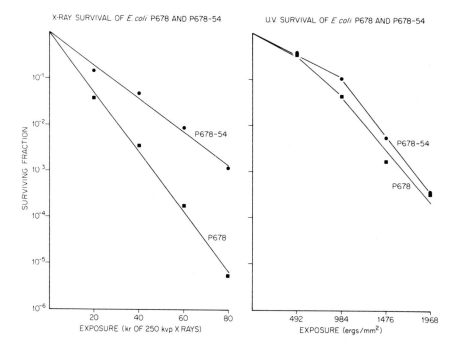

Fig. 2 - X-ray and ultraviolet (2650 Å) survival curves comparing the sensitivities of Escherichia coli P678 (■) and P678-54 (●), the minicell forming mutant. Cultures were grown in nutrient broth, irradiated in stationary phase of growth and plated on nutrient agar after irradiation.

MODIFICATIONS OF RADIOSENSITIVITY IN NUCLEATE AND ANUCLEATE AMOEBA FRAGMENTS

Yvette Škreb[x], Durda Horvat and Magda Eger

Institute for Medical Research, Zagreb, Yugoslavia

Summary

The objective of this study was to establish :
1. How similar are a real enucleation and the treatment of nucleate Amoeba fragments by actinomycin D or irradiation ?
2. What is the effect of actinomycin D, irradiation or both treatments on the nucleate and anucleate fragments ?
3. Whether radiation damage is reparable.

The selected parameters were the metabolism of RNA and proteins followed either by autoradiography after incorporation of specific radioactive precursors or by quantitative determination of these macromolecules.

It is concluded :
1. that the three treatments applied separately produce the same effect, and that in both nucleate and anucleate amoeba fragments exposed separately to the effect of radiation or actinomycin D the inhibition level of the synthesis is similar;
2. that after successive application of the two treatments the fragments react differently. Nucleate fragments are more affected by successive application of the two agents than anucleate ones by a single agent. However, the two agents applied to anucleate fragments produce a synergistic effect ;

---
[x] Head, Laboratory for Cell Biology.

3. that immediately after UV radiation, visible light as well as cooling induce restoration in both kinds of fragments. However, with time, restoration becomes more intense in nucleate fragments, while anucleate fragments die quickly.

The results are discussed in connection with cytoplasmic DNA.

Introduction

One of the aspects of molecular biology is the field of macromolecular interactions. Methods for analysing these interactions vary a great deal. In addition to biochemical analyses of homogenates whose shortcoming is to allow no observation of macromolecular interactions in the living cell, there are some methods less sensitive but fit to elucidate these intricate relationships.

It is customary to say that the most important part of a cell is the nucleus which owes its essential function to the presence of DNA. Although a cell without its nucleus is bound to perish, by comparing it with a nucleate cell of the same kind it is possible, on the one hand to deduce the role of the nucleus, and on the other, to see in which order a cytoplasm deprived of its nucleus loses its cellular functions.

An anucleate cell is a cell whose nucleus is eliminated either by micromanipulation or section. Thus, what remains in an anucleate cell is the cytoplasm and its inclusions, including the chondriome.

What is the purpose of the mechanical operation ? To prevent the new information from the nucleus to reaching the cytoplasm. Thus, the anucleate cell only contains the information it possessed at the moment of the section, which information depended on the length of the messenger-RNA life. However, the cell itself may be injured by the operation.

Besides this method, which is not always applicable, it is possible to damage nuclear structures by physical and chemical agents and in this way prevent the transmission of information. However, the agents used may act also directly on the cytoplasm. For this reason one must be very careful when speaking about a cell anucleated in this manner, because this is a cell whose nucleus is no longer functioning and whose cytoplasm is damaged.

Numerous problems related to anucleate cells have attracted the attention of a great many authors. But the cells convenient for such studies are far less numerous. Among them it is Amoeba that has become the subject of our interest.

Amoeba proteus is easy to cultivate. Its advantage is that it can be easily dissected in two parts by a glass needle under the binocular microscope. After a few hours the fragments obtained can be differenciated, the nucleate ones assuming the normal appearance of whole Amoeba and the anucleate ones remaining spherical (Fig. 1).

Brachet (1) was among the first to show that the anucleate Amoeba fragments can not only survive for at least two weeks but that they also possess a measurable metabolic activity. In this way he refuted a certain number of hypothesis which ascribed to the nucleus a primordial role for cytoplasmic activity.

Mazia and Hirschfield (2) have approached the radiobiology of anucleate Amoeba fragments by studying their survival following ultraviolet irradiation. They have shown them to be far more sensitive than the nucleate Amoeba fragments.

The famous experiment of Ord and Danielli (3) obtained by transplanting an irradiated nucleus into a non-irradiated cytoplasm, or a non-irradiated nucleus into an irradiated cytoplasm has revealed a particular radiosensitivity of the cytoplasm although the whole Amoeba is extremely radioresistant.

Mention should also be made of a great many studies by Plaut and co-workers (4,5) who tried to evaluate the functions of both the nucleus and the cytoplasm of Amoebae. Our own contribution consisted in trying to elucidate some functions of anucleate Amoeba fragments.

Among the many problems which arose when we started to work with Amoebae, we choose for this report the following three questions :
1. Is it possible to treat nucleate Amoeba fragments in such a way as to make them comparable to anucleate ones? Would it be possible to do so by using for instance, actinomycin D which can inhibit the expression of genetic

information ?
2. Do the actinomycin D - treated nucleate fragments react to irradiation in the same way as the anucleate fragments ?
3. Are the irradiated anucleate fragments just as susceptible to repair as the nucleate ones ?

Material and Methods

To answer these questions the following material and methods have been used :

The Amoebae, kept in pure culture and fed on Tetrahymena piriformis (ciliates), are allowed to fast for several days to eliminate nucleic acids and the proteins of the ciliates. In all the experiments with the fragments obtained by section, as already described, the fragments are starving because anucleate fragments are incapable of nourishing themselves.

The analysis of the survival of fragments after each experimental treatment allowed the choice of doses to be applied later. We choose doses of the agents which gave about the same survival. The parameters used were the metabolism of the RNA and proteins, that of the DNA representing some difficulties that we will explain later. The metabolism was on the one hand investigated by means of the incorporation of specific radioactive precursors. The percentage of incorporation was evaluated by the autoradiographic method. The counting of β traces under the microscope in the cytoplasmic or nuclear units allowed an adequately precise estimation of both the incorporation level and the effect of the treatment. On the other hand, according to the experimental conditions, the percentage of the RNA and proteins was evaluated by suitable micromethods. More details are given in previous papers (6,7).

Results

The results may be divided into two groups, the first one relating to the effect of actinomycin D and γ radiation on the groups of cells shown in table 1.

ANUCLEATE SYSTEMS: BACTERIA AND ANIMAL CELLS

Let us consider figure 2, which summarizes the autoradiographic results. The same marking in the columns represents the same treatment. In each case we can observe an inhibition of incorporation. The effect of the treatment with irradiation or actinomycin D on the nucleus is quite clear and more pronounced after the application of actinomycin D than after irradiation. This shows that by using these agents we have considerably damaged the nuclear metabolism.

What was the effect of the treatments on the cytoplasm?

As can be seen in Fig.2, the inihibition of the incorporation of cytidine is of the same order in nucleate cells treated with either one or the other agent and also in the anucleate control cells.

If looking now at the nucleate Amoeba fragments treated with both agents, it may be seen that they incorporate cytidine far less than do the anucleate cells treated with either one or the other agent but far more than the anucleate cells treated with both agents successively.

As regards the incorporation of phenylalanine, the inhibitions were practically of the same order of magnitude.

On the basis of these results, we may answer our first question affirmatively. The percentage of the inhibition of the RNA and protein synthesis has proved to be the same in nucleate fragments treated with either one or the other agent and in anucleate fragments.

As regards the second question, let us see table 2, which sums up these relations.

To simplify the estimation, in the first column on the left (1), the incorporation in nucleate fragments is taken to be 100 % and the percentage of remaining radioactivity is calculated in all treated nucleate as well as anucleate fragments. In the second column on the right (2), the incorporation in anucleate fragments is taken to be 100 %.

Thus an irradiated anucleate fragment incorporates less than an irradiated nucleate one, but the proportion between the treated cells and the control group is the same in both cases.

An irradiated anucleate cell incorporates twice as much as a nucleate cell treated with actinomycin D prior to irradiation. This shows that the effect of physical enucleation on the cytoplasm is smaller than the effect of the treatment with actinomycin D prior to irradiation of the nucleate cell.

To obtain a quasi-total inhibition of incorporation, the anucleate fragments should be treated with both actinomycin D and irradiation.

Can the anucleate fragments subjected to irradiation recover in the same way as the whole Amoebae or nucleate fragments?

Let us remember that recovery in whole Amoebae exists and that it is very intense in cases when they can be fed. However, we have to make comparison with anucleated fragments which cannot be nourished and therefore their recovery is pretty weak in a mineral medium, whereas in the dark it does not exist at all.

However, according to our experience gained long ago, in some special cases after UV irradiation, both nucleate and anucleate fragments may be photorestorated if subjected to the action of visible light (8). Not only did the survival prove better but also the percentage of RNA and protein in all the fragments increased rapidly.

In the present experiments, the second part of our work, the fragments were first treated with UV irradiation, then they were exposed to the action of low temperature (6°C) in the dark for two hours. The effect of these conditions on whole Amoebae allowed a good recovery.

The percentage of RNA and proteins in the fragments was evaluated by micromethods.

The experimental groups were as shown in table 3.

Preliminarily the estimation of RNA and proteins were performed in the cooled control fragments and they practically showed no difference as compared with the controls kept at normal temperature, 20°C.

Fig.3 summarizes the results obtained. First of all it may be seen that an irradiated nucleate cell practically shows the same percentage of RNA and proteins as the anucleate cell. This means that we could answer our first question affirmatively. However, after 24 h the irradiated nucleate cell has a somewhat higher amount of RNA, but a somewhat lower amount of proteins.

As regards the second question, the initial differences between nucleate and anucleate fragments remained the same in the course of the first 24 h.

As regards the recovery after cooling, it proved very clear in nucleate fragments, the percentage of the RNA and protein almost reaching the controls, while in anucleate fragments it was of the same order of magnitude than if they were only irradiated.

The percentages are summarized in table 4. These are roughly the results obtained, which allow us to answer the three essential questions posed at the beginning. However, these answers give rise to several other problems.

Discussion

We will try to discuss the following problems :
1. How similar are a real enucleation and the treatment of nucleate fragments by actinomycin D or irradiation?
2. What is the effect of both treatments on the nucleate and anucleate fragments ?
3. Is there any correlation between the recovery by cooling or by visible light and the content of nuclear and cytoplasmic DNA.

1.- We have been able to conclude that, under our experimental conditions and with the parameters used, UV or γ irradiation or actinomycin D treatment produced the same effect as anucleation. This means that we have answered our first question affirmatively. However, if we now compare the effect of both agents on the nucleate fragments, on the one hand, and on the anucleate ones on the other, we may state that : the inhibition in all cases is almost of the same order of magnitude, which means that concerning immediate effects the presence or absence of the nucleus is not of essential importance for this kind of treatment.

What does the presence or the absence of a nucleus mean (9) ? Surely, it means the presence or absence of the DNA. However, it must not be forgotten that our Amoebae have a considerable amount of cytoplasmic DNA, its content being 20 times as high as the content of the nuclear DNA, as we were able to measure (10) by the very refined fluorometric technique of Kissane and Robins. Consequently, if actinomycin D binds to the guanine of the nuclear DNA, it is rather likely that it also binds to the guanine of the cytoplasmic DNA. This is what might explain the important inhibitory effect on the anucleate fragments and the slight differences between the effects on the two types of fragments. It is also well known that even the RNA of anucleate cells is affected by actinomycin D (11).

The same related to the effects of irradiation which are of the same order of magnitude irrespective of whether the fragments are nucleated or anucleated.

2.- Regarding the action on the cell of the two agents applied successively, we known :
a) that there is an additive effect if the total effect is equal or inferior to the sum of the effects of the two agents applied separately ;
b) that a synergistic effect takes place if the final effect is superior to the sum of the two agents applied separately ;
c) that a sensitisation takes place if one of the two agents does not act or acts little by itself but increases the effect of the other.

Thus, if a nucleate cell treated with actinomycin D and irradiation is compared with an anucleate cell treated with irradiation or actinomycin D, the incorporation in the former is half the incorporation of precursors in the latter. In the case of the nucleate cell there is an addition and in the case of an anucleate cell this effect is only partially additive.

However, in the case of a double-treated anucleate cell there occurs synergy and the inhibition of incorporation is total. A few hours after treatment the divergence between the two kinds of fragments is greater if complex treatments are applied. It is possible in certain cases to follow the modalities of recovery during 24 h.

3.- A few years ago, a very clear-cut photoreactivation was observed in the two types of fragments. However, the classical restoration, in the mineral medium in the dark, seems to take place only in nucleate fragments and is stimulated by cooling.

Here again the role of the nuclear and the cytoplasmic DNA should be evaluated. It is only the abundance of the cytoplasmic DNA that gives ground to the supposition that it plays a role which is not always the same as the role of the nuclear DNA. Does it play a part in photorestoration?

Jagger, Prescott and Gaulden (12) have recently confirmed our results but they are not conclusive as to whether the photoreactivation is connected to the DNA or possibly also to the RNA.

Concerning the effect of cooling, we could only assume that the enzymatic systems which take part in this restoration are those connected also with the repair of the nuclear DNA.

Analizing the parallelism between nucleate and anucleate Amoeba fragments we may conclude as follows (Table 5):
1. All the three treatments if applied separately give similar effects;
2. If nucleate and anucleate Amoeba fragments are submitted to actinomycin D or irradiation separately, they react by a similar inhibition of the macromolecular synthesis (RNA and proteins);
3. If they are exposed to the two agents successively, they react differently : the nucleate cells are more affected by the two agents applied successively than the anucleate cells treated with one agent only, but the two agents produce a total synergy in anucleate cells ;
4. There is no marked difference between nucleate and anucleate cells regarding recovery immediately after UV irradiation. Cooling as well as visible light seem to stimulate recovery of the two types of fragments. However, with time the differences become more pronounced. While the recovery in nucleated fragments is greater, the anucleate fragments perished. In other words, a biological and radiobiological equivalent in nucleate and anucleate cells can be obtained with chemical and physical agents only if the use of these agents and the duration of the experiments are carefully choosen.

Acknowledgments

The authors are indebted to Dr. M.Drakulic and Prof. M.Errera for helpful criticism.

They are indebted to Bernarda Lončar, Jadranka Makvić and Desanka Margeta for technical assistance.

This work was made possible thanks to the financial

support offered by the Federal Council for Scientific Research and Council for Scientific Research of Croatia.

References

1. Brachet,J. (1955) - Biochim.Biophys.Acta, 18, 247.
2. Mazia,D. and Hirshfield,H.J. (1950) - Exptl.Cell.Res., 2, 58.
3. Ord,M.J. and Danielli,J.F. (1956) - Quarterly J.Micr. Science, 97, 29.
4. Wolstenholme,D.R. and Plaut,W. (1964) - J.Cell Biol., 22, 505.
5. Plaut,W. and Rustad,R.C., (1959) - Biochim.Biophys.Acta, 33, 59.
6. Škreb,Y. and Errera,M. (1957) - Exptl.Cell Res., 12,649.
7. Škreb,Y. and Bevilacqua,Lj. (1962 a) - Biochim.Biophys. acta, 55, 250.
8. Škreb,Y. and Bevilacqua,Lj. (1962 b) - Exptl.Cell.Res., 28, 430.
9. Keck,K. (1969) - Int.Rev.Cytol., 26, 191.
10. Benzinger,Lj., M.Sc.thesis, University of Zagreb (1966)
11. Škreb,Y., Eger,M. and Horvat,D. (1967) - Biochem.J., 105, 53.
12. Jagger,J., Prescott,D.M. and Gaulden,M.E. (1969) - Exptl.Cell Res., 58, 35.

ANUCLEATE SYSTEMS: BACTERIA AND ANIMAL CELLS

Table 1 - Treatments of Amoeba fragments

1. Nucleate Amoeba fragments - control group
2. Anucleate Amoeba fragments
3. Nucleate Amoeba fragments treated with actinomycin D, 50 µg/ml, for 2 hours
4. Nucleate Amoeba fragments treated with γ rays, 100 Krads
5. Nucleate fragments treated first with actinomycin D and then with γ rays
6. Anucleate Amoeba fragments treated with actinomycin D, 50 µg/ml, for 2 hours
7. Anucleate Amoeba fragments treated with γ rays, 100 Krads
8. Anucleate fragments treated with actinomycin D and γ rays

Table 2 - Percent of incorporation of radioactive precursors in Amoeba fragments after treatments

	^{3}H-cy-tidine		^{14}C-phenyl-alanine	
	(1)	(2)	(1)	(2)
1. Nucleate Amoeba fragments	100%		100 %	
2. Nucleate fragments + actinomycin D	60		55	
3. Nucleate fragments + γ rays	54		68	
4. Nucleate fragments + actinomycin D + γ rays	18		18	
5. Anucleate Amoeba fragments	60	100%	64	100%
6. Anucleate fragments + actinomycin D	40	66	36	56
7. Anucleate fragments + γ rays	37	61	52	80
8. Anucleate fragments + actinomycin D + γ rays	-	-	-	-

Table 3 - Treatments of Amoeba fragments

1. Non-treated nucleate Amoeba fragments as control group
2. Nucleate Amoeba fragments irradiated with UV 2400 ergs/mm^2
3. Irradiated nucleate fragments exposed to the cold 6°C, for 2 hours
4. Anucleate Amoeba fragments
5. Anucleate Amoeba fragments UV irradiated
6. Anucleate Amoeba fragments irradiated and then exposed to the cold

Table 4 - Percent RNA and protein content remaining in Amoeba fragments after treatments

	RNA		Proteins	
	2 h	24 h	2 h	24 h
1. Nucleate fragments	100%	84%	100%	50%
2. Nucleate + UV	72	51	87	31
3. Nucleate + UV + cooling	83	80	94	41
4. Anucleate fragments	70	42	77	46
5. Anucleate + UV	55	38	68	20
6. Anucleate + UV + cooling	62	30	68	20

Table 5 - Final effects of treatments on inhibition of RNA and protein synthesis in _Amoeba_ fragments

1. Nucleate _Amoeba_ fragments + act.D	as enucleation
2. Nucleate fragments + γ rays or UV	as enucleation
3. Nucleate fragments + act.D + γ rays	additive effect
4. Anucleate fragments + act.D	partial additive effect
5. Anucleate fragments + γ rays or UV	partial additive effect
6. Anucleate fragments + act.D + γ rays	synergistic effect
7. Nucleate cells + UV + visible light or cooling	recovery
8. Anucleate cells + UV + visible light	photorecovery
9. Anucleate cells + UV + cooling	no recovery after 2 h

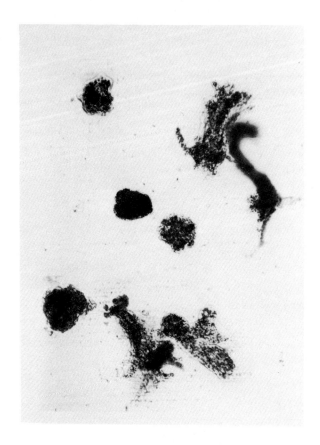

Fig. 1 - Microphotography showing nucleate and anucleate Amoeba fragments.

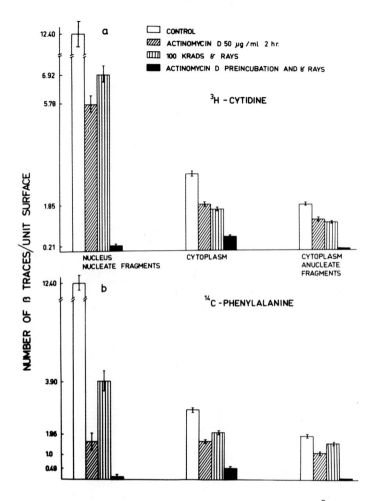

Fig. 2 - Effect of treatments on the uptake of ^{3}H-cytidine (a) and ^{14}C-phenylalanine (b) by Amoeba fragments. The numbers represent the mean of grain count per 1 nuclear or cytoplasmic unit $\pm$ standard error.

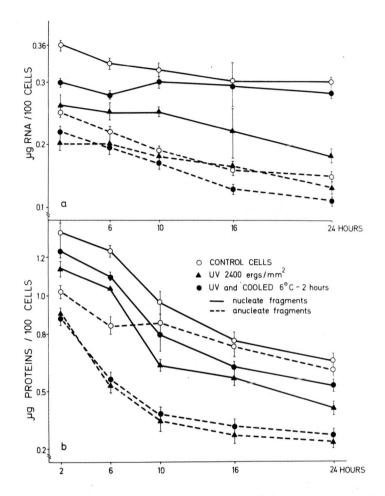

Fig. 3 - Effect of treatments on RNA (a) and protein (b) estimation, in μg per 100 Amoeba fragments. The numbers represent the mean of 8 experiments ± standard error.

SESSION 2
Chairman: Zénon M. Bacq

HETEROGENEITY OF MEMBRANE-BOUND POLYRIBOSOMES OF MOUSE MYELOMA CELLS IN TISSUE CULTURE

M. Zauderer and C. Baglioni

Department of Biology
Massachusetts Institute of Technology
Cambridge, Massachusetts 02139, U.S.A.

Introduction

Protein synthesis in animal cells is carried out by polyribosomes that are either free in the cytoplasm or that are associated with membranes of the endoplasmic reticulum. These two classes of polyribosomes synthesize different proteins (1-5) : the free polyribosomes synthesize proteins that remain in the cytoplasm or migrate to the nucleus, whereas membrane-bound polyribosomes synthesize proteins that are secreted by the cell and other proteins. The nature of the non-secretory proteins synthesized by membrane-bound polyribosomes has not been clarified yet ; it can only be inferred that such proteins are synthesized on membrane-bound polyribosomes by the observation that cells that do not secrete protein have membrane-bound polyribosomes active in protein synthesis (6-7).

We have been studying the biosynthesis of immunoglobulins in mouse myeloma cells in tissue culture. These cells are specialized in the synthesis of a single type of immunoglobulin, different for different cell lines, that is secreted in the culture medium. Approximately 10-15 % of the protein synthesized by exponentially growing cells is immunoglobulin ; non-dividing cells synthesize an even larger proportion of immunoglobulin. These cells are extremely stable in tissue culture and have been cultured over three years without any change in their biosynthetic activity.

Our aim in these studies has been to define the membrane-bound polyribosomes that are involved in immunoglobulin synthesis and to investigate the mechanism by which a class of messenger RNA (mRNA), that specifying proteins to be secreted by the cell, becomes localized on membrane-bound polyribosomes. The results obtained suggest that membrane-bound polyribosomes are heterogeneous and that two different classes of polyribosomes are associated with membranes of the endoplasmic reticulum. Approximately half of the polyribosomes can be released from membranes by treatment with salt solutions, whereas the other polyribosomes can only be released by detergents. This indicates that polyribosomes of the two classes are associated with membranes in a completely different way ; those polyribosomes that can only be released by detergent are involved in the biosynthesis of immunoglobulins.

Materials and methods

Cells. Mouse myeloma cells of the P-3 line were obtained from Dr.M.Cohn of the Salk Institute to whom we are deeply grateful for help and advice. The cells were cultured in roller bottles in suspension. Dulbecco's medium (GIBCO, New York) supplemented with 10 % horse serum was used. The cells were harvested at a concentration of approximately 5×10^5 cells/ml, when they were still growing exponentially. The cells were concentrated for incubation with radioactive isotopes by centrifugation for 5 min at 1,000 g. The cells were then resuspended in the same medium at a concentration of $1 - 2 \times 10^6$ cells/ml and incubated in a rotary shaker under 5 % CO_2.

Cell fractionation. The cells were collected by centrifugation and resuspended in 1 ml of RSB solution (0.01 M Tris/HCl pH 7.4 and 1.5 mM $MgCl_2$). The cells were homogenized by 15 strokes of a loose fitting Dounce homogenizer and the nuclei were spun down in 5 min at 1,000 g. Membrane-bound ribosomes were obtained from the postnuclear supernatant by either method $\neq$ 1 : 5 min centrifugation at 27,000 g (8) ; or by method $\neq$ 2 : 30 min centrifugation at 80,000 g over a 15-30 % sucrose gradient in RSB (7).

The pellet obtained by method ≠ 1 was resuspended in 5 ml
of RSB and centrifuged again 5 min at 27,000 g to eliminate contamination with free polyribosomes. The pellets obtained were resuspended in 0.5 ml of RSB by stirring slowly with a magnetic stirrer. All the operations described
have been carried out at 0-4°C.

Gradient analysis. Polyribosomes were analyzed by centrifugation for 18 h at 16,000 rpm on 15-50 % sucrose density
in RSB. The absorbancy at 260 nm was monitored in a Cilford
recording spectrophotometer. The fractions collected were
precipitated with 5 % trichloroacetic acid, filtered and
counted in a liquid scintillation counter.

Analysis of nascent chains : The P-3 cells were labeled
5 min with [35S] methionine after being transferred to a
methionine-less medium. Polyribosomes were obtained by centrifugation of the different fractions (see text) for 2 h
at 50,000 rpm on 15 % sucrose. The pellets obtained were
resuspended in RSB and digested for 1 h at 37°C with
50 μg/ml of pancreatic ribonuclease. An antiserum prepared
with purified P-3 immunoglobulin (a kind gift of Dr.Donato
Cioli and David Schubert of the Salk Institute) was then
added, followed after 1 h by goat anti-rabbit immunoglobulin serum. The equivalence between the rabbit anti-P-3 serum and the goat anti-rabbit was established by standard
immunological techniques (9). The immune precipitate obtained after 24 h at 2 - 4°C was dissolved in a buffer containing sodium dodecyl sulphate (SDS) and analyzed by
acrylamide gel electrophoresis according to Maizel (10).
The gels were sliced for counting.

Results

Approximately 25 % of the ribosomes of P-3 cells are
associated with membranes of the endoplasmic reticulum(8).
Incubation of the cells with agents that inhibit protein
synthesis and cause polyribosome dissociation leads to the
release of approximately 50 % of the membrane-associated
ribosomes into the pool of free ribosomes (Fig.1). This
has been shown with three different inhibitors of protein
synthesis, pactamycin, pederine and NaF, that inhibit pro-

tein synthesis by preventing translocation of the initiator tRNA from the aminoacyl ribosomal site to the peptidyl site (11). Pactamycin and pederine inhibit specifically initiation only when used at very low concentrations ; these compounds are bound by ribosomes and the inhibition of protein synthesis is thus irreversible. The inhibition by NaF can be reversed by washing the cells and incubating them with fresh medium (12). We have previously shown that when polyribosomes reform, a fraction of free polyribosomes, roughly equivalent to that which remains bound to the membranes when polyribosomes disaggregate, becomes associated with the membranes of the endoplasmic reticulum (12).

Two classes of membrane-bound ribosomes can thus be distinguished : those that are released from membranes upon polyribosomes disaggregation and those that remain associated with membranes and can only be released by the solubilizing action of detergents. These results are in substantial agreement with those of other investigators (6,7,13) that have observed an heterogeneity of membrane-bound ribosomes by treating the membranes with EDTA (6,13) or ribonuclease (7) or the intact cells with puromycin(7). The nature of this heterogeneity has so far not been clarified, nor have the products of the two classes of polyribosomes been investigated.

Sabatini and collaborators (14) have reported that concentrated solutions of KCl release a fraction of membrane-bound ribosomes from the membranes of the endoplasmic reticulum. We have applied their observation to the study of the membrane-bound ribosomes of mouse myeloma cells in tissue culture, with the aim of establishing the nature of the biosynthetic products of the two classes of ribosomes. In a preliminary investigation, preparations of membrane-associated polyribosomes were obtained by method ≠ 1 (see Methods) ; the pellet obtained was resuspended in RSB and increasing volumes of 2 M KCl, 8 mM $MgCl_2$ and 50 mM Tris/HCl pH 7.5 were added. The membranes were then reisolated and the amount of ribosomes associated with them determined (Fig.2). A progressive release of ribosomes from membranes of cells that had been treated with inhibitors of initiation was observed, whereas only approximately half of the ribosomes of control untreated cells were released from membranes by KCl solutions(Fig.2).

ANUCLEATE SYSTEMS: BACTERIA AND ANIMAL CELLS

The release of these ribosomes occurs at rather low KCl concentrations and is almost complete at 0.4 M KCl, 2 mM $MgCl_2$. This KCl concentration was thus used in all the following experiments directed at studying the nature of the ribosomes released and at indentifying their products.

An analysis of the free ribosomes, of the ribosomes released by 0.4 M KCl and of the ribosomes released by detergent is shown in Fig.3. Polyribosomes active in protein synthesis are present in all fractions ; pulse-labeling experiments have shown that the polyribosomes become labeled at approximately the same rate. The pattern of 80S ribosomes and of ribosomal subunits is however quite different in different fractions. Free 40S subunits are only present in the free ribosome compartment, together with 60S subunits ; free 60S subunits are found in the ribosome fraction released from membranes by KCl. In this fraction there is also relative to the polyribosome peak a large amount of 80S. In the ribosome fraction released by detergent there is little 80S and practically no 60S ribosomal subunit. Two small peaks of ribonucleoprotein particles that appear in the top part of the gradient have been identified with the two subunits of mitochondrial ribosomes (15). A control experiment was carried out to eliminate the possibility that the detergent used to release membrane-bound ribosomes could change the pattern of ribosomal subunits. Detergent was thus added to an aliquot of the preparation of free ribosomes and the pattern analyzed (Fig. 3C). No change in the relative amount of ribosomal subunits was observed. This indicated that the presence of ribosomal subunits and 80S ribosomes among the different fractions most likely reflects the real distribution of these organelles among different cellular compartments.

We have studied the product of the different classes of polyribosomes by isolating the nascent chains after pulse-labeling with $[^{35}S]$ methionine. A rabbit antiserum directed against the immunoglobulin secreted by P-3 cells was then used to combine with immunoglobulin nascent chains and a goat anti-rabbit immunoglobulin antiserum added to precipitate the antigen-antibody complexes formed (see Methods). This indirect precipitation method is useful to select peptide chains that react with the antiserum but gives a somewhat high background. It was thus necessary to analyze the labeled protein present in the precipitate by acrylamide gel electrophoresis. This analysis showed the presence of peptide chains that migrate close to the position of authentic L and H chains of P-3 immunoglobulin in the nascent chains obtained from the polyribosomes released by detergent but not in the polyribosomes released by KCl or in the free polyribosomes (Fig.4).

The presence among the nascent chains of peptide chains that migrate almost like L and H chains may seem surprising but one has to keep in mind that these chains have been selected by the immune precipitation ; it seems likely that the antiserum reacts preferentially with nascent chains that are almost complete and possess thus the antigenic determinants that the antibody recognizes. A control experiment was carried out to exclude the possibility that newly synthesized immunoglobulin contaminates specifically the polyribosome fraction released by detergent. The cells were labeled 3 min with [^{35}S] methionine and then incubated for 2 min with 0.1 mM puromycin. This antibiotic is known to cause premature termination of nascent chains that are thus released from polyribosomes (16). It should not have any effect on a contaminant protein, which is not bound to ribosomes via tRNA. The polyribosomes were isolated from the puromycin-treated cells and the immune precipitation carried out. The labeled protein recovered was a small fraction of that recovered from cells not treated with puromycin and when this protein was analyzed by acrylamide gel electrophoresis we failed to observe the H and L chains peaks.

Discussion

The experiments reported indicate that ribosomes are associated with membranes of the endoplasmic reticulum in myeloma cells in at least two different ways. Approximately half of the ribosomes are loosely associated. These ribosomes are released from membranes when the cells are incubated with inhibitors of initiation of protein synthesis, and conversely become associated to membranes when polyribosomes reform. The other half of the ribosomes are tightly associated and can be released from membranes by detergents or by high concentrations of KCl after polyribosome run off. Only these tightly associated ribosomes synthesize immunoglobulin.

Our data suggest that the tightly bound ribosomes are specialized in the synthesis of proteins secreted into the cisternae of the endoplasmic reticulum. These proteins can be secreted by the cell after undergoing some further processing in the Golgi apparatus (17), or they can possibly remain associated with the cell membrane. It has been recently shown (18) that cell-bound immunoglobulins (immunoglobulins bound to the surface of human lymphoma cells) are secreted into the cisternae of the endoplasmic reticulum and then transported to the cell surface, but not secreted outside the cell. This might explain the presence of tightly bound ribosomes in cells that have no secretory function (7). Some components of the cell membrane are possibly secreted into the cisternae and then transported to the cell surface, where they are assembled into plasma membranes.

The tightly bound ribosomes are intimately associated with the membranes of the endoplasmic reticulum. We have made some preliminary observations by electron microscopy of endoplasmic reticulum preparations washed with 0.4 M KCl to remove the loosely bound ribosomes (Fig.5B). It is quite clear that the ribosomes still associated with membranes after washing with KCl are those lining the outside of the cisternae. The interaction of these ribosomes with the membrane is stabilized by the presence of peptidyl-tRNA. This acts as an anchor (19) preventing release of ribosomes from membranes under conditions that are effective in dissociating 80S ribosomes, like M KCl. 60S ribosomal subunits that become associated to membranes during the assembly of membrane-bound ribosomes are released by 0.4 M KCl (Fig.3), since no peptidyl-tRNA is associated with them. It seems however likely that addition of a 40S subunit to form an 80S ribosome makes the binding to the membrane tighter since higher concentrations of KCl are required to release the 80S ribosomes from membranes than to release 60S ribosomal subunits (Fig.2).

The mechanism by which KCl promotes dissociation of tightly bound ribosomes from membranes has not been investigated in detail. High concentrations of KCl have been used to extract proteins from cell membranes (20). It seems likely that under these ionic conditions protein-protein interactions are weakened and dissociation of membrane components is promoted. We have previously postulated that assembly of membrane-bound polyribosomes occurs through the binding of 60S ribosomal subunits to sites on membranes that are complementary to 60S ribosomal subunits (15); membrane and ribosomal proteins are presumably involved in this interaction, the strength of which is presumably dependent on ionic conditions. Several ribosomal proteins are removed from ribosomes by 0.5 M KCl (21) ; it seems likely that removal of other ribosomal or membrane proteins by more concentrated KCl solutions is responsible for the dissociation of tightly bound ribosomes from membranes.

The loosely associated ribosomes are already released at 0.4 M KCl. This indicates that their nascent polypeptide chains are not secreted through the cisternae of the endoplasmic reticulum (in this case the peptidyl-tRNA would act as an anchor and prevent ribosome dissociation from membranes). It has not been established, however, whether the loosely associated ribosomes are associated with membranes of the endoplasmic reticulum in a specific way. It seems possible that they interact with the membranes through their nascent polypeptide chains. This is in agreement with the finding that inhibitors of initiation prevent the synthesis of peptide chains and cause the release of the loosely associated ribosomes from membranes. A similar class of membrane-associated ribosomes exists in HeLa cells ; release of nascent chains by puromycin in these cells promotes dissociation of these ribosomes from membranes (7).

The postulated interaction of loosely bound ribosomes with membranes through their nascent chains may account for the release of this class of ribosomes by 0.4 M KCl or by a short ribonuclease treatment (7). At this relatively high KCl concentration relatively weak protein-protein interactions may be abolished. Moreover, only nascent chains that are sufficiently long and have acquired some three-dimensional structure may interact with membranes. A polyribosome may thus be associated to membranes through the nascent chain of one or a few ribosomes. When mRNA is hydrolyzed by RNase the other ribosomes are released from membranes.

We can thus provide a rational explanation for the experimental observations on the ribosomes loosely associated with membranes. There is no direct evidence, however, to prove that this explanation, although rational, is the correct one. The major objection concerns the specificity of the association between loosely bound ribosomes and membranes. It has not been excluded that these ribosomes are "contaminants" that become associated with the endoplasmic reticulum during or after the cells are broken and fractionated. Sections of our endoplasmic reticulum preparations show groups of ribosomes that are not in close association with membranes (Fig.5A) ; these ribosomes are removed by washing with 0.4 M KCl, the procedure that we have used to separate the loosely associated ribosomes from membranes. These ribosomes do not seem to lay thus in close contact with membranes.

Decisive evidence for a specific role of loosely associated ribosomes may possibly come from a study of their biosynthetic products, if the proteins that they synthesize differ from those synthesized by either free or tightly bound ribosomes. It is tempting to speculate that these proteins may be membrane components ; if these proteins associate with membranes with high affinity, this might explain the behaviour of this class of polyribosomes. We are currently investigating these possibilities with the aim of clarifying the function of the two classes of ribosomes bound to membranes.

Acknowledgements

This investigation has been supported by Grant AI08116 of the National Institutes of Health. We are grateful to Mrs.Rebecca Hinderlie and to Mr.Ira Goldberg for their help and to Dr.Michael Rosbach for discussions and criticism.

References

1. Redman,C.M., Siekevitz,P. and Palade,G.E. (1966) - J.Biol.Chem., 241, 1150.
2. Takagi,M. and Ogata,K. (1968) - Biochim.Biophys.Res.Commun., 33, 55.
3. Redman,C.M. (1968) - Biochim.Biophys.Res.Commun., 31, 845.
4. Hicks,S.J., Drysdale,J.W. and Munro,S.N. (1969) - Science, 164, 584.
5. Ganoza,M.C. and Williams,C.A. (1969) - Proc.U.S.Nat.Acad.Sci., 63, 1370.
6. Attardi,B., Cravioto,G. and Attardi,G.E. (1969) - J.Mol.Biol., 44, 47.
7. Rosbash,M. and Penman,S. (1971) - J.Mol.Biol., 59, 227.
8. Kimmel,C.B. (1969) - Biochim.Biophys.Acta, 182, 361.
9. Kabat,E.A. and Mayer,M.M. (1961) - Experimental Immunochemistry, Thomas : Springfield, Illinois.
10. Maizel,J.V. (1966) - Science, 151, 988.
11. Baglioni,C., Jacobs-Lorena,M. and Meade,H. in preparation.
12. Bleiberg,I., Zauderer,M. and Baglioni,C. submitted for publication.
13. Sabatini,D.D., Tashiro,Y. and Palade,G.E. (1966) - J.Mol.Biol., 19, 503.
14. Sabatini,D.D., Blobel,G., Nonomuza,Y. and Adelman,M.R. (1969) - in "Proceedings of the First International Symposium on Cell Biology and Cytopharmacology" (F.Clementi and E.Trabucchi, Eds.) Raven Press : New York.
15. Baglioni,C., Bleiberg,I. and Zauderer,M. (1971) - Nature, New Biol., 232, 8.
16. Nathans,D. (1964) - Proc.Nat.Acad.Sci.U.S.A., 52, 585.
17. Schenkein,I. and Uhr,J. (1970) - J.Cell Biol., 46, 42.
18. Sherr,C.J. and Uhr,J.W. (1971) - J.Exp.Med., 133, 901.
19. Sabatini,D. and Blobel,G. (1970) - J.Cell Biol., 45, 146.
20. Reisfeld,R.A. and Kahan,B. (1970) - Fed.Proc., 29, 2034.
21. Miller,R.L. and Schweet,R. (1968) - Arch.Biochem.Biophys., 125, 632.
22. Jacobs-Lorena,M., Brega,A. and Baglioni,C. (1971) - Biochim.Biophys.Acta, 240, 263.

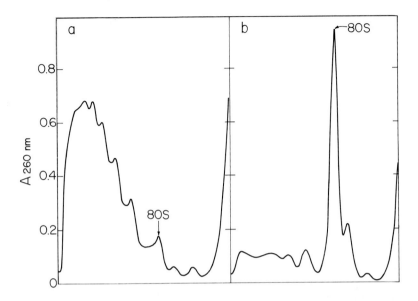

Fig. 1 - Effect of an inhibitor of initiation on membrane-bound polyribosomes. Two 150 ml cultures of P-3 cells (2×10^5 cells/ml) uniformly labeled with [^{14}C] uridine (see Methods) were used, one as a control, whereas the other was incubated 20 min with 10 mg/ml of pederine (22). The cells were collected by centrifugation and the membrane-bound ribosomes prepared by method ≠ 1 (see Methods). They were analyzed by centrifugation for 18 h at 16,000 rpm on 15-50 % sucrose gradients in RSB. The $A_{260\ nm}$ was analyzed in a recording spectrophotometer. A, control ; B, incubated with pederine. The fractions collected were counted to determine the relative proportion of membrane-bound ribosomes. The cells incubated with pederine had 57 % of the membrane-bound ribosomes present in control cells.

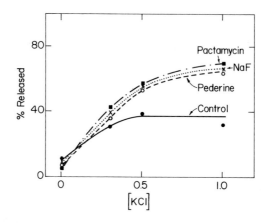

Fig. 2 - Release of ribosomes from membranes by KCl. Four 100 ml cultures of P-3 cells uniformly labeled with [^{14}C] uridine were incubated in the following way :
1. control unincubated;
2. incubated 20 min with 4×10^{-7} M pactamycin (11) ;
3. incubated 20 min with 10 ng/ml of pederine
4. incubated 60 min with 15 mM NaF.

The cells were collected by centrifugation and endoplasmic reticulum preparations obtained by method $\neq$ 1 (see Methods). The membrane pellets were resuspended into 0.5 ml of RSB and subdivided into 4 aliquots to which increasing amounts of 2 M KCl, 8 mM MgCl$_2$ and 50 mM Tris/HCl pH 7.5 were added to obtain the KCl concentrations reported in the figure. The membranes were reisolated by centrifugation for 5 min at 15,000 rpm and the proportion of ribosomes released determined. The % ribosomes released by KCl is reported. The cells incubated with pactamycin, pederine and NaF had respectively 63%, 46% and 55% of the membrane-bound ribosomes present in control cells.

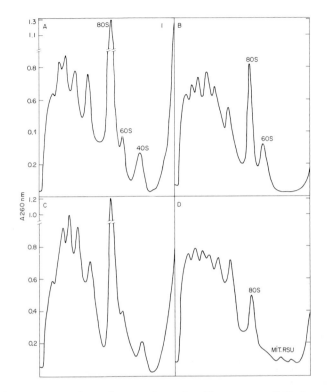

Fig. 3 - Free and membrane-bound ribosomes of P-3 cells. A 300 ml culture of P-3 cells was used to obtain by method ≠ 1 (see Methods) post-mitochondrial supernatant and a pellet containing endoplasmic reticulum. The pellet was resuspended into 0.5 ml of RSB and 0.5 ml of 0.4 M KCl, 2 mM $MgCl_2$, were added. The membranes were reisolated by 5 min centrifugation at 15,000 rpm and the supernatant applied to a sucrose gradient (B). The pellet obtained was resuspended in 0.5 ml of RSB and 0.5 ml of detergent added (see Methods). The analysis of this fraction is shown in D. The postmitochondrial supernatant was analyzed directly (A) or after addition of an equal amount of detergent (C). The position of the 80S, 60S and 40S peaks is indicated in A, B and C; the position of mitochondrial ribosomal subunits is indicated in D.

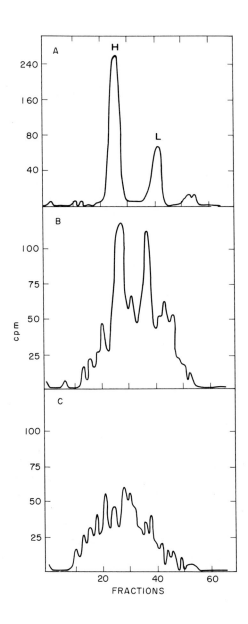

Fig. 4 - Acrylamide gel electrophoresis of the nascent chains isolated by immune precipitation from loosely membrane-associated and from tightly associated ribosomes. P-3 cells grown in complete Dulbecco's medium were collected by centrifugation and transferred to methionine-free medium. The cells were equilibrated with this medium for a few minutes and then [^{35}S] methionine was added. After 5 min the incubation was stopped by the addition of cold saline and the cells collected by centrifugation. A pellet containing endoplasmic reticulum was then obtained by method $\neq$ 2 (see Methods). The pellet was resuspended into 0.5 ml of RSB and 0.5 ml of 0.4 M KCl, 2 mM MgCl$_2$ added. The membranes are reisolated by 5 min centrifugation at 15,000 rpm. The supernatant and the membranes dissolved with detergent were applied over a layer of 15 % sucrose in RSB and spun at 50,000 rpm for 1 h to pellet polyribosomes. These were resuspended in RSB and digested 30 min at 37°C with 1 mg/ml of pancreatic ribonuclease. Rabbit anti-P-3 antiserum diluted 1 : 10 was then added and after 24 h at 4°C goat anti-rabbit immunoglobulin antiserum. The conditions for optimal precipitation of rabbit immunoglobulins had been previously determined by standard immunological techniques (9). The immune-precipitate obtained was redissolved in the SDS-containing buffer used for gel acrylamide electrophoresis and analyzed according to the Method of Maizel (10). The gels were sliced and counted in a liquid scintillation counter. Uniformly labeled P-3 immunoglobulin was prepared from the culture medium of P-3 cells incubated with [^{3}H] proline, an amino acid that is not present in the tissue culture medium. The immunoglobulin was isolated by ammonium sulphate precipitation between 30% and 55% saturation. The separation of H and L chains by gel electrophoresis is shown in A. The analysis of the immune-precipitate of nascent chains obtained from tightly-bound polyribosomes is shown in B and that of loosely-bound polyribosomes in C.

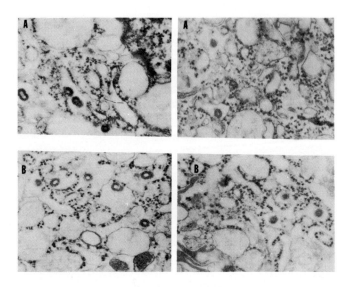

Fig. 5 - Sections of untreated endoplasmic reticulum preparations (A) and of endoplasmic reticulum treated with 0.4M KCl (B). The pellet containing endoplasmic reticulum was obtained from postnuclear supernatant in RSB (A) or from postnuclear supernatant brought to a 0.4M KCl concentration (B) by method #2 (see Methods). The pellets were fixed 60 min in 2% gluteraldehyde in 0.1M Na-cacodylate buffer pH 7.4, washed twice in this buffer and fixed for 2 hours with 2% osmium tetroxide. The pellets were then washed with saline, stained with 0.5% Mg-uranyl acetate, dehydrated, embedded and sliced.

9 S HAEMOGLOBIN MESSENGER RNA FROM RETICULOCYTES
AND ITS ASSAY IN LIVING FROG CELLS

C.D.Lane, G.Marbaix and J.B.Gurdon

Department of Zoology, South Parks Road, Oxford, England

and

Department of Molecular Biology, University of Brussels,
Rhode-St.-Genèse, Belgium

Introduction

Mammalian reticulocytes are cells which have undergone physiological enucleation; they may be expelled into the peripheral circulation under condition of acute anaemia. However, these cells are still able to synthesize protein (95 % of which is haemoglobin) for several hours. This observation suggests that there is a stable haemoglobin messenger RNA in the cytoplasm of the reticulocyte. Over the years several different research groups have attempted to isolate the messenger RNA for haemoglobin : for the reticulocyte system presents a favourable opportunity, perhaps the most favourable, for isolating a messenger RNA corresponding to a well characterized protein.

In Chantrenne's laboratory, Marbaix and Burny (1964) (1), Burny and Marbaix (1965) (2), Marbaix, Burny, Huez and Chantrenne (1966) (3), Chantrenne, Burny and Marbaix (1967) (4) and Huez, Burny, Marbaix and Lebleu (1967)(5) were able to isolate and characterize a 9 s RNA fraction from rabbit reticulocyte polyribosomes which had many of the properties expected of the messenger RNA for globin chains.

In 1969, Lockard and Lingrel (6) showed that a 9 s RNA fraction obtained from mouse reticulocyte polyribosomes by a method similar to that used in Chantrenne's laboratory directs the synthesis of mouse globin β chains in a rabbit reticulocyte system. Further experiments showed that this fraction also contained the message for the α chain (7). It was thus proved that the 9 s RNA fraction from reticulocyte polyribosomes contains the messages for both globin chains.

In this communication a method for preparing highly purified 9 s RNA is described (8) : it is an improvement on that described by Huez et al., (5). The activity of the 9 s RNA is assayed using living frog cells, namely Xenopus oocytes ; the advantages of this new assay system for messenger RNA are discussed.

Preparation of 9 s RNA from rabbit reticulocytes

Rabbits were made anaemic and were bled as described previously (1). Blood cells were collected by low-speed centrifugation and washed twice in an isotonic saline medium. Differential lysis of reticulocytes and leucocytes was brought about by adding two volumes of cold hypotonic saline, shaking for two minutes and then re-establishing isotonicity by adding 0.2 volumes of 1.5 M NaCl. The lysis of leucocytes was thus avoided. Intact cells and stroma were eliminated by low-speed centrifugation. The stroma-free lysate was layered on 5 ml cushions of 30 % sucrose made in 5 mM Tris-HCl, 20 mM KCl, 1 mM Mg acetate, pH 7.6 and the tubes were centrifuged for two hours at 30,000 rpm. in the type 30 rotor of the Spinco ultracentrifuge.

The polyribosomes pellets were then resuspended in 6.6 mM phosphate buffer containing 33 mM EDTA (pH = 7.0) and 1 ml fractions of the resulting suspension (10 mg/ml) were layered on 15-30 % linear sucrose gradients made in 0.01 M sodium phosphate pH 7.0. After a 40 h centrifugation at 25,000 rpm and 4°C in the SW 27 Spinco rotor, fractions were collected and analyzed for absorbance at 260 nm.

Fig.1 shows the result of such a centrifugation. Thus treatment with 33 mM EDTA causes complete disaggregation of the polyribosomal structure : the ribosomal subunits fall apart, and the 5 s RNA and the 9 s messenger RNA are released as ribonucleoproteins sedimenting at 8 and 14 s respectively (8). As shown in Fig.1 the centrifugation step was long enough to cause the ribosomal subunits to reach the bottom of the tube, thereby achieving good resolution in the 14 s mRNP region of the gradient.

The sucrose gradient fractions which contain the mRNP were pooled and the solution was made 0.4 M in NaCl and 1 % (w/v) in sodium dodecylsulphate (SDS). Two volumes of ethanol were added and RNA and protein were allowed to precipitate by standing overnight at -20°C. The precipitate was collected by low-speed centrifugation and was dissolved in a solution of 1 % (w/v) SDS (1 ml of solution per 100-200 µg of RNA). 1 ml portions of the resulting RNA solution were layered on 10-20 % sucrose gradients made in 10 mM Tris-HCl, pH 7.4. After centrifugation (using a Spinco SW 27 rotor) for 40 h at 25,000 rpm and at 4°C, fractions were collected and analyzed for their absorbance at 260 nm. Fig.2 shows the result of such a centrifugation step. The 9 s RNA is now almost pure and is devoid of any contaminating 5 s RNA. Fractions corresponding to the pure 9 s RNA were pooled and were made 0.4 M in NaCl. Two volumes of ethanol were added and the RNA was allowed to precipitate at -20°C overnight. The 9 s RNA was then collected by centrifugation, dissolved at a concentration of 100 µg/ml, in 10 mM Tris-HCl buffer pH 7.4 and residual proteins were removed by two extractions with a mixture of chloroform and isoamyl alcohol (24:1 v/v). The RNA solution was then dialysed for 15 h against two changes of 1,000 volumes of double-distilled water, and was then lyophilized.

Starting with 10 rabbits (having a reticulocytosis of from 50 to 90 %) one can obtain 1 mg of purified 9 s m-RNA.

The assay of globin messenger RNA in frog oocytes

A sensitive assay system for messenger RNA has recently been developed by Lane, Marbaix and Gurdon (9) and by

Gurdon et al. (10). The following description illustrates the use of this system for the assay of haemoglobin m-RNA.

In the experiments described here, oocytes were taken from frogs (Xenopus laevis) which had been induced by hormone treatment to ovulate between two and four weeks previously. This was done to ensure that the larger oocytes were actively growing. Oocytes were injected dry and without removal of follicle cells (11). Micropipettes of 10-15 µ diameter were calibrated to deliver a volume of 50-70mµl the actual amount being kept constant in each related series of experiments. All samples to be injected were taken up in or dialysed into the following injection medium: 88 mM NaCl, 1.0 mM KCl, 15 mM Tris-HCl, pH 7.6. Injected cells were incubated at 19°C in the saline medium described by Gurdon (11). Oocytes were labelled by incubation in this culture medium containing from 0.1 to 1 mCi/ml of ^{3}H-histidine (30-50 Ci/mM). At the end of the incubation cells were frozen.

Using these techniques, the following experiments were carried out to test the effects of introducing 9 s reticulocyte RNA into living oocytes.

30 oocytes were each injected with the standard solution containing haemin (0.5 mg/ml) and 9 s RNA (700 µg/ml) and were incubated for 6 h in a ^{3}H-histidine-containing medium. Another batch of 30 cells were treated similarly except that the injectate contained no 9 s RNA. After freezing, cells were homogenized in a 0.0522 M glycine-0.0522 M Tris - 0.1 % (w/v) histidine buffer (pH 8.9) containing 5 mg/ml of commercial rabbit haemoglobin. Homogenates were centrifuged at 75,000 g for 30 minutes at 4°C and the resulting supernatant was applied directly to a 140 x 1 cm G100 Sephadex column equilibrated with 0.0522 M glycine - 0.0522 M Tris buffer (pH 8.9). Fig. 3 shows the elution profiles obtained from the samples described. The presence of 9 s RNA in the injectate is linked to the appearance of a peak of radioactivity with an elution profile identical to that of added marker rabbit haemoglobin (9).

Material from the haemoglobin region of the Sephadex profile was mixed with the appropriate solutions and loaded onto a polyacrylamide gel. After electrophoresis using the method of Moss and Ingram (1968) (12), a sharp peak of radioactive material was found to have moved with the marker haemoglobin (Fig.4). In controls lacking 9 s RNA material taken from the optical density peak of haemoglobin showed no preferential migration of radioactivity with marker haemoglobin (9). Thus the molecules whose synthesis is caused by the presence of 9 s RNA are, under these conditions, indistinguishable from haemoglobin, in overall charge distribution as well as in size.

If the material whose synthesis within the frog oocyte is directed by rabbit 9 s RNA is really haemoglobin, it should be possible to show that this material contains α and β globin chains. Thus the haemoglobin region of the Sephadex elution profile was pooled, mixed with more marker rabbit haemoglobin and treated with acid acetone to yield a precipitate of globin. If this protein preparation is then analysed by chromatography on a CM-cellulose column (after Dintzis,1961) (13), it is found that the presence of 9 s RNA in the injection medium is linked to the appearance of radioactive material that co-chromatographs with marker rabbit globin chains (Fig.5). If 9 s RNA is omitted from the injection mixture, no material which preferentially migrates with globin chains can be detected (9).

Definite identification of the polypeptides peculiar to 9 s RNA injected oocytes was provided by an analysis of the tryptic peptides derived from this material. α and β chains, uniformly labelled with ^{14}C-histidine, were prepared by incubating rabbit reticulocytes with this aminoacid; the separated chains were then mixed with oocyte-derived ^{3}H-histidine labelled α and β chains. These mixtures were digested with trypsin and the resulting α and β tryptic peptides were analyzed using a cation exchange resin column. Fig.6 shows the result of such an analysis for the tryptic peptides of the β chain. It is clearly seen that all the peptides from the two different sources coincide perfectly, both qualitatively and quantitatively (except for one peptide in the latter instance).

It is worth mentioning that neither 26 s, 18 s and 4 + 5 s RNA from the reticulocyte nor 9 s RNA from another tissue (mouse myeloma) cause the synthesis of globin in injected oocytes. Furthermore, the translation of 9 s RNA within the oocyte does not require added haemin : nor does this process require any reticulocyte-specific factors (10).

The globin synthesized resembles rabbit as opposed to frog globin, as shown by its chromatographic behaviour on CM-cellulose : the frog proteins elute after both rabbit chains. Moreover, the sequence of the β chain of Rana esculenta globin is known to be very different to that of the rabbit β chain (14). Although the sequence of the β chain of Xenopus globin is not known, one can, nonetheless, argue that it is likely to be quite different from that of the rabbit β chain. Consequently, Xenopus globin tryptic peptides would not be expected to coincide with tryptic peptides derived from rabbit globin.

Conclusions and Discussion

The results obtained confirm those of Lockard and Lingrel (1969) (6) who used an in vitro system to provide evidence that the 9 s RNA fraction from reticulocytes contained haemoglobin messenger activity. However, this work presents the most complete characterization of α and β globin chains synthesized, in a heterologous system, under the direction of 9 s messenger RNA.

A more interesting point is that the oocyte provides an excellent system for testing messenger RNA's (16). Firstly, the translation of the message occurs in a normal living cell and is, therefore, less likely to be affected by artefacts than it would be in a cell-free system. Secondly, injected oocytes are able to translate messages very efficiently, for long periods of time : for example, one can obtain, over a 24 h period, as much as 10^6 dpm of labelled Hb from 1 mµg of injected m-RNA ; furthermore, ribosomal RNA is not inhibitory to this process, for 9 s RNA mixed with more than 10 times as much 18 s RNA is still translated efficiently. Lastly, Xenopus oocytes appear to show very little species specificity with regard to the type of m-RNA which they can translate. Other types of m-RNA's have already been tested in this system, both 9 s RNA from mouse reticulocytes (15) and 9 s m-RNA from mouse myeloma cells (16) are readily translated.

ANUCLEATE SYSTEMS: BACTERIA AND ANIMAL CELLS

Acknowledgements

This work was supported by the Medical Research Council. G. Marbaix is "Chargé de Recherches" of the Belgian "F.N.R.S." and part of his contribution to this work has been made possible by an EMBO fellowship to work at Oxford. The method of preparation of 9 s RNA described in this communication has been devised by A. Burny, G. Huez, B. Lebleu and G. Marbaix.

References

1. G. Marbaix and A. Burny, Biochem. Biophys. Res. Commun., 16, 522 (1964).
2. A. Burny and G. Marbaix, Biochim. Biophys. Acta, 103, 409 (1965).
3. G. Marbaix, A. Burny, G. Huez and H. Chantrenne, Biochim. Biophys. Acta, 114, 404 (1966).
4. H. Chantrenne, A. Burny and G. Marbaix, Progr. Nucl. Acid. Res. Mol. Biol., 7, 173 (1967).
5. G. Huez, A. Burny, G. Marbaix and B. Lebleu, Biochim. Biophys. Acta, 145, 629 (1967).
6. R. E. Lockard and J. B. Lingrel, Biochem. Biophys. Res. Commun., 37, 204 (1969).
7. J. B. Lingrel, in "Methods in Protein Synthesis", Ed. A. E. Laskin and J. A. Last, Marcel Dekker, Inc. N.Y. in press
8. B. Lebleu, G. Marbaix, G. Huez, A. Burny, J. Temmerman and H. Chantrenne, Europ. J. Biochem., 19, 264 (1971).
9. C. D. Lane, G. Marbaix and J. B. Gurdon, J. Mol. Biol., in press
10. J. B. Gurdon, C. D. Lane, H. R. Woodland and G. Marbaix, Nature, London, in press.
11. J. B. Gurdon, J. Embryol. Exp. Morph., 20, 401 (1968).
12. B. Moss and V. M. Ingram, J. Mol. Biol., 32, 481 (1968).
13. H. M. Dintzis, Proc. Natl. Acad. Sci. U.S.A., 47, 247 (1969).
14. J. Chauvet and R. Acher, FEBS Letters, 10, 136 (1970).

15. C. Drewienkiewicz and C.D. Lane, unpublished observations.

16. J. Stavnezer, J.B. Gurdon, R.C. Huang and C.D. Lane, in preparation.

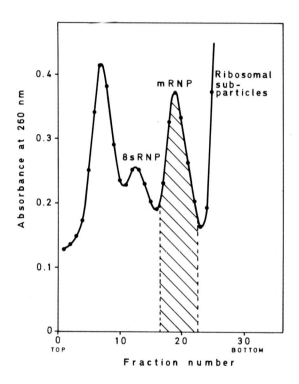

Fig. 1 - The figure shows the sedimentation pattern obtained by centrifuging a suspension of reticulocyte polyribosomes treated with 33 mM EDTA. The ribosomal subunits have reached the bottom of the tube. The 15-30 % linear sucrose gradients used were made up in 10 mM sodium phosphate buffer pH 7.0 : centrifugation was performed at 6°C for 40 h at 25,000 rpm. using a Spinco SW 27 rotor.

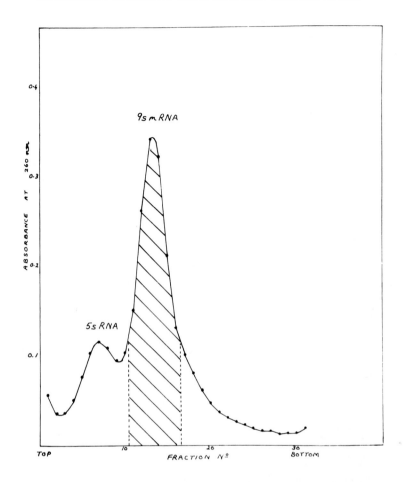

Fig. 2 - The figure shows the sedimentation pattern of the RNA extracted from the ribonucleoprotein material corresponding to the hatched area shown in Fig.1. The 10-20 % linear sucrose gradients used were made up in 5mM Tris-HCl pH 7.4 : centrifugation was performed at 4°C for 40 h at 24,000 rpm using the Spinco SW 25.1 rotor.

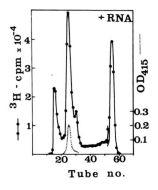

Fig. 3.a shows the Sephadex G 100 elution profile of supernatant material from oocytes which were injected with 9 s RNA (dissolved in a haemin-containing buffer at a concentration of 700 mg/ml) and cultured for 6 h in ^{3}H-histidine at 1 mCi/ml.

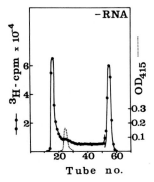

Fig. 3.b is identical except that the injectate contained no 9 s RNA. Cts/min refer to material from a single oocyte Open (O——O) refer to optical density at 415 nm and closed circles (●——●) refer to cts/min.

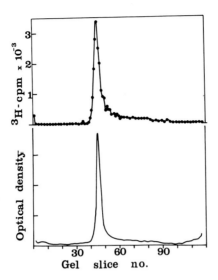

Fig. 4 - This figure shows the distribution pattern of material from the haemoglobin region of the Sephadex elution profile after gel electrophoresis at 5 mA/gel for 90 min.
The continuous line (———) refers to optical density at 415 nm and the closed circles (●———●) to cts/min.

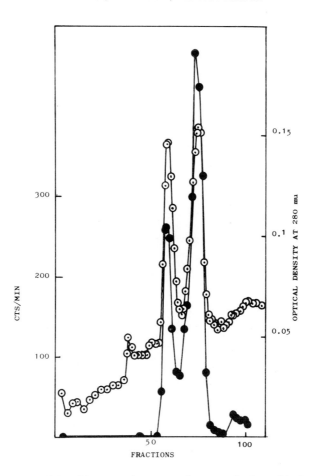

Fig. 5 - This figure shows carboxymethyl-cellulose chromatography of globin chains. Material from the haemoglobin region of the Sephadex elution profile was used to prepare globin ; further marker rabbit haemoglobin was added prior to this stage.
Open circles (O———O) refer to optical density at 280 nm and closed circles (●———●) refer to cts/min.

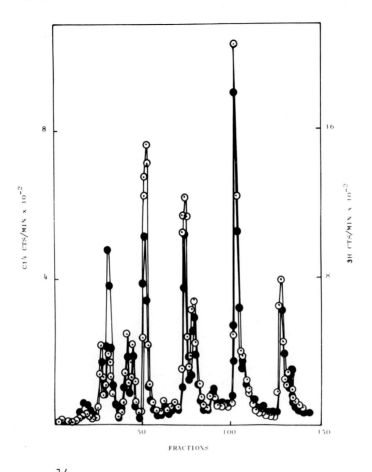

Fig. 6 - ^{14}C-His-β globin chains from rabbit reticulocytes were mixed with ^{3}H-His-β globin chains from oocytes and the resulting solution digested with trypsin. The peptide mixture was then analysed on a Technicon Chromobead (type P) cation exchange resin eluted with a formic acid-pyridine gradient, thereby yielding the pattern shown. Open circles (O———O) refer to ^{3}H cts/min and closed circles (●———●) refer to ^{14}C cts/min.

SESSION 3
Chairman: Maurice Errera

THE SQUID GIANT AXON : A SIMPLE SYSTEM FOR THE STUDY OF MACROMOLECULAR SYNTHESIS IN NEURONAL FIBRES

by Antonio Giuditta

International Institute of Genetics and Biophysics,
Via Marconi 10, Naples, Italy

The shape and structure of neurons are characterized by the presence of axonal processes which may extend for considerable distances from the cell body. This unusual cellular design has long posed the problem of the origin of the axonal components, particularly of those which are synthesized in subcellular organelles patently absent from the axon. This is the case of proteins synthesized on ribosomal structures which have not been identified in axons (1), and should also be the case of axonal RNA (2,3).

The establishment of the concept of axonal flow (4) and more recently the finding of a rapid transport of proteins toward the most distal regions of the axon (5) have provided a reasonable explanation to this problem. According to this view, proteins and other biochemical components synthesized in the neuronal cell body are constantly moving toward the nerve ending at rates varying from a few mm/day to two orders of magnitude higher. The nature of the components transported by the slow and by the fast flows and the mechanisms subserving them have not been fully elucidated.

Axonal proteins, however, may also be synthesized in the nerve fibre itself. The first indication of this phenomenon came from a study of the kinetics of recovery of acetylcholinesterase activity in nerves after its irreversible inactivation by diisopropylfluorophosphate. The enzyme reappeared uniformely along the entire nerve segment, showing no sign of a proximo-distal gradient as would have been

expected on the basis of its transport from the cell bodies
(6). In later experiments it was shown that isolated nerves
(without their cell bodies) were capable of incorporating
radioactive amino acids into the protein fraction of the
axons (7-11).

The main technical problem in this type of work consists
in discriminating the contribution of the axon itself from
that of the surrounding tissue, in the first place myelin
and gliacells. In most cases such a separation has been
accomplished with autoradiographic (11) or microdissection
techniques (7,10). While the former method does not allow
the biochemical analysis of the labelled axonal components,
the applications of the latter technique appear limited by
the small size of the dissected samples. From this point of
view, the giant axon of the squid is endowed with several
distinct advantages for studies on the synthesis of proteins
and other molecules in neuronal fibres. In the first place,
it is a single fibre resulting from the fusion of several
hundreds axons whose cell bodies are located in a special
lobe of the stellate ganglion (12). Its unusually large diameter (up to 0.5-1.0 mm in some species) and the length of
the segment which can readily be dissected (5-6 cm) allows
one to obtain a sample of pure axoplasm large enough for
biochemical analysis. In addition, the giant axon lends itself easily to electrophysiological stimulation in vitro
(it requires only sea water as its bathing medium), thus
allowing the study of the effect of activity on several biochemical parameters. In this search the biochemist can make
use of the large amount of neurophysiological information
accumulated on this system which for its simplicity has long
represented for neurophysiologists the analogue of what viruses have been for molecular biologists.

Our first experiments with the giant fibre were aimed at
showing the occurrence of protein synthesis in the absence
of the neuronal cell bodies and were carried out with the
species of squid (Loligo pealii) available at the Marine
Biological Laboratory in Woods Hole, Mass. (9). The two most
medial stellate nerves of the animals were tied at both ends
with silk threads, dissected out of the mantle and the giant
axons were carefully freed from the surrounding small fibers
and connective tissue. The clean axons were then incubated
at 18° - 20° in Millipore-filtered sea water containing 5mM
Tris pH 7.8 and 10 μc/ml of a mixture of 15 ^{14}C-labelled
amino acids (New England Nuclear Co.). After different times

of incubation the axons were washed with water, blotted on tissue paper and the axoplasm was separated from the sheath with the help of a roller. The axonal sheath is a composite structure which cannot be equated with the glial compartment since it contains also the axonal plasma membrane with some adhering axoplasm, connective tissue and a few fragments of small fibres not fully discarded during the dissection. In both samples analysis of the radioactivity was carried out on the TCA-soluble and protein fractions prepared according to standard procedures.

Under these conditions labelled amino acids entered readily the soluble pools of the axoplasm and of the sheath with rates which were linear during the first hour or so but tended to decrease thereafter. An approximately linear increase lasting several hours was found for the radioactivity of the protein fractions obtained from the axoplasm and from the axonal sheath. The specific activity of the latter compartment was at least ten times higher than that of the axoplasmic protein. Treatment of the axon with 200 µg/ml puromycin or cycloheximide inhibited the incorporation of radioactive amino acids into the axoplasmic protein more than 90%, while chloramphenicol at the same concentration was somewhat less effective, producing an inhibition of approximately 70%. Axoplasmic protein could therefore be synthesized in the giant fibre itself, under conditions which excluded the contribution of the cell bodies. Similar conclusions were independently reached by other investigators with giant axons isolated from the chilean squid (Dosidicus gigas) (8) and were subsequently confirmed by us with squids available at the Zoological Station of Naples (Loligo vulgaris).

The occurrence of protein synthesis in neuronal fibres and in nerve endings (13,14) poses several problems which relate ultimately to the functional role of this process. Two of the most pressing ones concern the nature and location of the synthetic apparatus and the nature of the synthesized protein species. Axonal proteins formed in situ might in principle be synthesized in the axon itself, either in the mitochondria or in other as yet unknown structures, or be transported in the axon from a synthetic site located in nearby structures, most likely glia cells. The second possibility entails an exchange of macromolecules across cellular boundaries. We have examined the occurrence of the last process in squid giant axons, by taking advantage of the possibility to obtain from them uncontaminated samples of axoplasm

(15). Axons from Loligo vulgaris were tied at both ends, freed from small fibers and connective tissue and incubated at 18° - 20° in Millipore-filtered sea water containing ^{125}I-albumin from human serum (Radiochemical Centre, Amersham; 2.5 μc/mg in concentrations up to 10 mg/ml. After different periods of time the axon was washed in a large volume of cold sea water and the axoplasm was extruded.

Radioactivity started to appear in the protein fraction of the axoplasm after a few minutes and kept increasing for approximately one hour, but decreased thereafter. On the other hand, a significant amount of radioactivity started to appear in the TCA-soluble fraction of the axoplasm only after approximately two hours, in concomitance with the decreased labelling of the protein fraction and presumably as result of its hydrolysis. The amount of ^{125}I-albumin entering the axon after 60 min was considerably higher at a protein concentration of 10 mg/ml than at 4 mg/ml. When the axon was incubated at 3° with 4 mg/ml radioactive protein, the amount of radioactivity which entered the axoplasm was insignificant. Control experiments indicated that the radioactivity found in the protein fraction of the axoplasm was indeed due to ^{125}I-albumin. The presence of this protein was detected by electrophoresis on polyacrylamide gel, which showed a radioactive profile identical to that of a sample of ^{125}I-albumin mixed with unlabelled axoplasm. It appeared furthermore that damage of the axonal sheath during the dissection could not be held responsible for the appearance of the radioactive protein in the axoplasm, since similar results were obtained after incubation of whole stellate nerves in vitro or after injection of ^{125}I-albumin in the proximity of stellate nerves in living squids. In both cases there was little or no handling of the giant fibre. It was also found that puncturing the axon with a needle did not increase the amount of radioactive protein recovered in the axoplasm in comparison with the contralateral undamaged axon. As result of these experiments we concluded that proteins of the size of human serum albumin were actively taken in by the giant axon at the approximate rate of 0.013 μg/mm^2/hr at 18° - 20°C. Although indicating that the neuronal plasma membrane can be crossed by macromolecules, these findings cannot be taken to suggest a glial site for the peripheral system of protein synthesis except in a circumstantial and indirect way. The problem of the nature and cellular location of this system is still to be elucidated.

Only preliminary information is available regarding the nature of the axoplasmic proteins synthesized in the giant fibre. Since most axonal proteins appear to be derived from the cell body (5) it might seem reasonable to expect that only a few discrete species are synthesized in situ. Electrophoretic analysis on polyacrylamide gel of the labelled axoplasmic proteins obtained from squid giant axons after an incubation of two hours with a mixture of ^{14}C-amino acids has revealed however that the radioactivity was distributed along the whole gel, in association with several and perhaps with all the protein bands (16). A result which appears to contradict such an expectation.

Labelled axoplasmic proteins are largely soluble. After incorporation times of one-two hours more than 90% of the protein-bound radioactivity is found in the supernatant fraction (25.000 x g for 20 min.), while in the axonal sheath only 40% of the radioactive proteins are soluble (9). With shorter periods of incubation the proportion of labelled axoplasmic protein found in the supernatant fraction appears to decrease, although the relatively low levels of radioactivity prevailing at these times make these determinations somewhat difficult (17). The finding that most of the axoplasmic proteins synthesized in the giant fibre are soluble has a bearing on the identification of the synthetic structure, which from this point of view does not seem to be identifiable with the mitochondrion since the protein synthesized in this organelle is largely if not exclusively membrane-bound. Neural mitochondria may however be different in this respect from those of other cellular types, since their protein synthetic apparatus has been found not to share the sensitivity to chloramphenicol common to conventional mitochondria. On the other hand, neural mitochondria have been reported to be inhibited by cycloheximide derivatives, which are typical inhibitors of the cytoplasmic ribosomal system (18-20). If these findings are correct (21) they would provide strong evidence for the occurrence of an unusual ribosomal system from which it would not be unreasonable to expect the synthesis of soluble proteins.

As mentioned previously, the giant axon of the squid represents perhaps the simplest system in which the biochemical aspects of nerve activity can be investigated. By stimulating giant axons with electrical square pulses above threshold we have examined the effects of action potentials on protein synthesis and on the uptake of ^{125}I-albumin utilizing the contralateral fibre as control. The conduction of action potentials was monitored on an oscilloscope screen. In the case of protein synthesis the experimental nerve was initially incubated for one hour with ^{3}H-leucine (Radiochemical Centre, Amersham; 5 c/mmole; 50 µc/ml) to insure labelling of the intracellular soluble pool and was then stimulated for two hours at 5 impulses/sec while being left in the same solution. The control nerve was treated similarly but was not stimulated. In eight experiments the specific radioactivity of the protein fraction of the stimulated giant axon was found to be 1.72 ± 0.2 times higher than that of the control. The difference was statistically significant with P<0.01. Other authors had reported an essentially similar effect in the giant axon of the chilean squid (<u>Dosidicus gigas</u>) stimulated for 10 min at 100 impulses/sec and incubated with a microinjected radioactive amino acid for 100 min since the start of the stimulation (8).

In the experiment with ^{125}I-albumin we used essentially similar conditions, except that the stimulation was carried out during the one hour of incubation with the radioactive protein (4 mg/ml). Analysis of the protein-bound radioactivity of the extruded axoplasm carried out in six experiments indicated that the amount of ^{125}I-albumin which had entered the stimulated axon was 2.4 ± 0.2 times higher than in the control fibre. The difference was statistically significant (P<0.01). This result is perhaps the best demonstration that the entrance of protein in the axoplasm requires intact and functioning axonal membrane and is not due to diffusion through damaged sites of the axonal sheath. The increase in protein uptake by a neuronal structure as result of activity appears to be relevant from the point of view of the mechanism involved and, more generally, in regard to the postulated exchange of macromolecules between glia and neurons which appears also to be markedly enhanced by activity (22,23). It is certainly in remarkable accord with the similar increase in protein synthesis brought about in the giant axon by electrophysiological stimulation.

References

1. Palay,S.L. et al. (1968) - J.Cell Biol. **38**, 193.
2. Edström,A. (1964) - J.Neurochem. **11**, 309.
3. Andersson,E. et al. (1970) - Acta Physiol.Scand. **78**, 491.
4. Weiss,P. et al. (1945)- Amer.J.Physiol. **143**, 521.
5. Lasek,R.J. (1970) - Int.Rev.Neurobiol. **13**, 289.
6. Koenig,E. and Koelle,G.B. (1961) - J.Neurochem. **8**,169.
7. Koenig,E. (1967) - J.Neurochem. **14**, 437.
8. Fischer,S. and Litvak,S. (1967) - J.Cell Physiol., **70**, 69.
9. Giuditta,A. et al. (1968) - Proc.Natl.Acad.Sci.US **59**, 1284.
10. Edström,A. and Sjöstrand,J. (1969) - J.Neurochem. **16**, 67.
11. Caston,J.D. and Singer,M. (1969) - J.Neurochem. **16**, 1309.
12. Bullock,T.H. and Horridge,G.A. (1965) - "Structure and function in the nervous system of invertebrates" (Freeman,W.H. & C°.) Vol.II,p.1481.
13. Morgan,I.G. and Austin,L. (1968) - J.Neurochem. **15**,41.
14. Autilio,L.A. et al. (1968) - Biochem. **7**, 2615.
15. Giuditta,A. et al. (1971) - Nature New Biology **229**,29.
16. Giuditta,A. (1970) - "Biochemistry of simple neuronal models", Ed. E.Costa and E.Giacobini, Raven Press, p.339.
17. Giuditta,A. and Dettbarn,W.D. - Unpublished experiments.
18. Gordon,M.W. and Deanin,G.G. (1968) - J.Biol.Chem. **243**, 4222.
19. Morgan,I.G. (1970) - FEBS Letters **10**, 273.
20. Haldar,D. (1971) - Biochem.Biophys.Res.Commun. **42**, 899.
21. Mahler,H.R. et al. (1971) - FEBS Letters **42**, 384.
22. Hyden,H. and Pigon,A. (1960) - J.Neurochem. **6**, 57.
23. Pevzner,L.Z. (1971) - J.Neurochem. **18**, 895.

ANUCLEATE STENTORS :
MORPHOGENETIC AND BEHAVIORAL CAPABILITIES

Vance Tartar

Department of Zoology, University of Washington
Seattle, USA

Stentor, a ciliate, is quite different from the other select unicellular organisms in which it has proved possible to produce anucleate systems routinely. There is a highly complex pattern of ectoplasmic organelles and the nucleus is segmented in the larger species. In the largest and best known, <u>Stentor coeruleus</u> (Fig.1) about 100 kineties or rows of body cilia with their fibrous interconnectives run from pole to pole, separated by bands of blue-green pigment granules which visibly and beautifully outline the pattern of the cell cortex. This pattern is not only polarized but also shows a circular anisotropy. In the sector below the mouthparts the kineties are closest together and the pigment bands visibly narrowest, increasing in width around the cell. The widest bands come to meet the narrowest in a locus of stripe contrast which defines the primordium site where new ingestive structures always originate as an oral primordium. The ingestive apparatus itself is a complex organelle, consisting of a band of membranelles which propel food to the mothparts. At the other end the cell tapers as a posterior pole ending in a holdfast by which a stentor can voluntarily attach to the substratum or release itself and swim away.

Paper read by title

Methods

To remove the nucleus from a stentor the animal is placed in a viscous solution of polyethylene oxide in which it cannot feed and will digest and void pre-existing food vacuoles, leaving the cell clear. This substance causes less shedding of pigment granules which would obscure the architecture of the cell cortex, but the stringiness of this high polymer interferes with operating. The stentor is therefore transferred into a drop of methyl cellulose against a black background, illuminated by reflected light from an embryological lamp. With a glass needle, 2 incisions are cut through the cell following the chain of macronuclear nodes (Fig.2 A). The specimen is then opened like a clam shell and the nodes excised with minimum cytoplasm. When there is an opaque accumulation of glycogenoid reserve granules at the posterior pole, special care is taken to tease this region with the needle in order to locate and remove the last node. With the needle, the "clam-shell" is then "shut", which is to say that the cell is encouraged to heal in approximately its original form by guiding the cell parts with the operating needle as they fuse together again.

In this operation we take advantage of 3 aspects of the stentor nucleus. The composite macronuclear nodes appear opaque white in darkfield illumination, and even more brightly after exposure by the cutting. The chain macronucleus is always located as illustrated and there are no stray nodes to search for. Removal of the macronuclear nodes will also delete all or most of the many micronuclei which lie close to them ; but even if a few should be left in the cell, they can be ignored. Schwartz (8) showed that stentors with micronuclei only behaved as if anucleate, that in such specimens the micronuclei never transform into functional macronuclei, and that a macronucleus functions normally after all micronuclei have been removed. These determinations by this very careful worker effectively remove for us the problems of the dimorphic nucleus of Stentor. Therefore in place of macronucleus, which is the only functional nucleus except in conjugation, I will simply speak of the nucleus.

Alternatively, nucleate and anucleate pieces of stentor

ANUCLEATE SYSTEMS: BACTERIA AND ANIMAL CELLS

can be produced rapidly by a single transection (Fig.2 B), again taking advantage of the fixed,sub-cortical location of the chain of nuclear nodes. Important organelles are left in the anucleate piece, as the illustration shows.

When a stentor dies on a slide, the nucleus remains intact for 2 or 3 days, its nodes still brightly opaque white in darkfield illumination within the disintegrating cytoplasm. The absence of any bright node(s) at the conclusion of an anucleate experiment then confirms that enucleation was complete. Or the specimen can be squashed in a flattened drop and searched for nodes.

It should not be overlooked that the dissected naked nucleus can do nothing but dies at once without the protection of the cytoplasm. Nuclear nodes exposed to the medium for more than a minute were picked up and trapped inside split anucleates but they were never viable, the same as with Amoeba and other cells.

Microsurgical experiments with stentor are done by hand and I doubt that a micromanipulator would help. The training potential of anyone with manipulative aptitude is great and after a few weeks of patient practice stentors can be enucleated at the rate of one per minute. The legend for figure 2 describes enucleation in some detail because I hope others will not feel barred from using Stentor when it appears to be the most suitable animal.

Capabilities of anucleate Stentor coeruleus

Early investigators seem to have been chiefly surprised that anucleate cells and eggs could do anything at all, as Gruber in 1885 was astonished when his enucleated stentor blithely swam away. Surprise implies the contradiction of an hypothesis, and perhaps the assumption was that the nucleus acts directly and immediately on the cell ; therefore that after one removed the "heart" of the cell it should die at once and certainly not go on performing vital functions. We now know that the nucleus acts indirectly, as the source of specific macromolecular synthesis, by messengers delivered to the cytoplasm. The following is an account of the capabilities of anucleate Stentor coeruleus.

1. Survival

Anucleate stentors survive for an average of 5.5 days. Starved nucleates live for an average of 4.4 days, probably because they are more active, are "more alive" and do not deteriorate, as will soon be described for anucleates.

Starved animals maintain their oral and body cilia into death, though their movement is very slow toward the end. Food vacuoles are digested and nutritional reserves exhausted. The cytoplasm becomes entirely pellucid, so clear that all the nuclear nodes are plainly visible. Pigmentation is maintained, the cell decreases in size somewhat but it does not become smaller and smaller until tiny and no part of the nucleus is consumed as emergency nutrition.

In anucleate stentors (Fig.3), after 2 days, movement and ciliary beating is greatly slowed, and the pigment begins to fade. Mothparts regress by uncoiling. Cytoplasm becomes progressively opaque until it is milky-white in darkfield illumination. Specimens decrease to half the original volumne, but this may indicate a dehydration causing the opacity. They become spherical, then vesiculate. Oral cilia are gradually resorbed and then the body cilia, or at least the spheres show no potruding fuzziness of body cilia like stopped normal stentors. The longer oral cilia twitch and tremble up to their disappearance and apparently the body cilia also as the anucleate sphere rotates almost imperceptibly though the body cilia cannot be seen. Thus in the absence of the nucleus the cytoplasm becomes abnormal and cortical structures are dissolved. These degenerative changes in the last days of an enucleate are entirely reversible through the 4th day. By implanting several new nuclear nodes the cell can be completely revived and will clone. On days 5 and 6, if surviving, recovery is incomplete and death, though postponed, finally ensues (11). This chronology for viable survival of anucleate cytoplasm is precisely that for Amoeba (2).

Survival is not extended by increased size of the specimen. Anucleate stentors can be grafted by their exposed endoplasms as readily as nucleates. Fusion masses of 3 to 6 anucleates lived no longer than singlet anucleate controls. The same is not the case with nucleates, in which masses outlive singlets. It appears that the combined reserves of masses are used to promote survival if the nucleus is present, but that this effective utilization and conversion of substance for continued maintenance is not possible in anucleates.

Implanting nuclear nodes from a different species does not promote but even hinders survival. When S.polymorphus nucleus was engrafted into S.coeruleus after enucleation, the foreign nucleus survived in its strange cytoplasm, but disharmonies apparently added to, rather than subtracted from, the stresses in anucleates.

Symbiotic algae do not promote survival of anucleates. This was tested in S.polymorphus from which the macronuclear nodes were successfully removed in spite of the abundant Chlorella because the specimens tabulated never regenerated excised mouthparts and died. Survival times were even less than for S.coeruleus.

Fusing fresh, anucleate cytoplasm to anucleates after they had become opaque and motionless does not result in a complete and "magical" revival of the specimen. Old and new cytoplasms apparently affected each other about equally and the specimen merely improved temporarily and survived longer.

In summary, anucleate stentors as now prepared can survive for as long as one week, much longer than previously supposed, which means that during the first days they probably have enough maintenance vitality to exhibit short-term capabilities if these are possible in the absence of the nucleus. Comparison with starved stentors shows clearly that anucleates are not merely starving but undergo progressive deterioration of cytoplasmic structures. The cytoplasm seems to require continual input from the nucleus as if turnover continued, becoming withdrawal without replacement. This contribution from the nucleus is so specific (in an immunological sense) that a foreign nucleus can offer no generalized aid, nor can fresh cytoplasm or the assistance of algal symbionts substitute for it.

2. Behavior

For about 2 days the behavior of anucleate stentors is entirely normal. Swimming is vigorous, the cells contract and extend, membranelles start and stop and beat in rhythm. This is to be expected because the organelles mediating behavior and the means of their coordination reside on the cell surface and cortex, and the energy for activity should be supplied by the semiautonomous mitochondria. Anucleates attach and release the holdfast. For one day they can feed but not thereafter ; whether because the ciliary beat becomes weaker or because the mouthparts begin uncoiling toward regression remains uncertain. Eventually the anucleate becomes too feeble to exhibit any behavior.

3. Observations on metabolism

Digestion can be followed in anucleates fed the relatively large Paramecium bursaria, bright green with algal symbionts (Fig.3). There is an expected lag in enzymatic capacity. One or 2 food vacuoles may turn brown after one day, as normally, but most of the ingested paramecia are soon expelled, undigested, from the cell.

Carbohydrate reserve granules are concentrated at the posterior end of the cell and clearly visible in darkfield. Starved nucleated stentors use up these reserves promptly after one day. In anucleates the reserves visibly decrease and may even disappear entirely but much more slowly, over a period of 3 to 4 days.

4. Contractile vacuole

The incisions by which a stentor is enucleated do not come near the contractile vacuole (Fig.4). In the anucleate the vacuole continues pumping, as noticed long ago by Balbiani (1) and Prowazek (7). The rate eventually decreases, and later swelling and vesiculation of the specimen atest the loss of osmiotic regulation (Fig.3). Anucleates can regenerate a contractile vacuole, but only if part of the system of associated canals remains, otherwise not. Schwartz (8) had suspected this incapacity for complete neoformation. A fluid vacuole forms in the fragments but it never pulsates. Flickinger and Coss (3) found that anucleate Amoeba also could not regenerate a contractible vacuole de novo.

5. Morphogenesis

Anucleates can form a new tail pole and holdfast after the original has been excised. This is not much of a morphogenetic accomplishment, as indicated by the fact that the regeneration requires only one hour in either nucleates or anucleates. Existing structures are reshaped. Lateral striping draws together to heal the cut and a taper is somehow effected. Terminal ectoplasm is then drawn out into branching pseudopod-like extensions which are adhesive.

The pattern of the lateral striping was disarranged in anucleates. Their capacity to recover the normal pattern was poor. In contrast, nucleates can reorient a completely minced cortical pattern into parallel, pole-to-pole striping in one day (10). Excised heads-only can reduce the enormous disproportion between total oral structures and the tiny bit of cell-body remaining, achieving small normal stentors, but only if one or more macronuclear nodes are left in the fragment (11).

When two "heads" are present in the same specimen, as by fusion of 2 animals, the heads migrate together and fuse as one bistomial ingestive structure (Fig.5). This unifying involves selective resorption, first of the cortical kineties and pigment stripes separating the two heads, and second, of the "unnecessary" sections of the membranellar bands which come together and initially separate the 2 frontal fields. Fused pairs of anucleate stentors can accomplish this unification by highly selective resorption of cortical structures (Fig.5).

Anucleates cannot regenerate the oral structures, as normal animals do in about 10 h. Even completely starved nucleate stentors can, before they die, regenerate excised ingestive organelles complete, confirming that starvation alone is not what is wrong with anucleates. The inconspicuous beginnings of a primordium are not even initiated in anucleates. But if an oral anlage is already present it can develop at most two stages further after the nucleus has been removed. At what is called stage 4, the membranellar band is already fully formed, kinetosomal reproduction and major synthesis is over, and what remains are essentially intracellular movements by which the posterior end of the band coils and invaginates to produce the gullet, and the lateral anlage is moved forward to its final location at the anterior end. Therefore, it is not surprising that some stage-4 regenerators, enucleated, can complete the oral regeneration.

Like regenerators, reorganizers and dividers can also complete their program in the absence of the nucleus (Fig. 6). Specific resorption of the oral structures which are spontaneously replaced in reorganizers begins and is completed after the nucleus has been removed. In dividers, formation of a fission line by circumferential sectioning of longitudinal kineties and pigment bands, as well as complete constriction of daughter cells occur without delay after enucleation.

A perhaps greater demand upon anucleates occurs when the membranellar cilia are cast off by immersing the stentors for 1-2 minutes in 20 % sucrose (13). It is assumed that the basal parts -the membranellar kinetosomes and their rootlets- are left in the cell, because on longer treatments with more dilute sugar the entire band is shed and the strip it occupied is healed over completely by pigmented ectoplasm. At first, short sprouts of cilia appear, and finally the completed cilia which move and coordinate normally. There may be as many as 25,000 membranellar cilia which are the largest cilia in a stentor. Therefore their regeneration represents a considerable accomplishment for anucleates, indicating residual capacities for protein synthesis or a protein pool in reserve.

My studies have been restricted to observations on living stentors. They describe situations and designate time shedules for the application of electron microscopy and biochemical analyses that could yield further and more specific details where these would appear to be fruitful.

Anucleate analysis : Stentor systems approaching the anucleate

Separation of nucleus and cytoplasm obviously does not allow us to assess the role of each in the complete cell. Without cytoplasm the nucleus comes to nothing and immediately degenerates. What little it can do alone we may leave to the in vitro biochemist. Without a nucleus, animal or holozoic cells run down and soon die. DNA cannot "make the

ANUCLEATE SYSTEMS: BACTERIA AND ANIMAL CELLS

cell" nor can cytoplasm alone long survive.

This dilemna in the analysis should not conceal that each major component of the cell, nucleus and cytoplasm with their enclosing membranes, have distinct roles in the life of the cell and organism. Were this not so, there would be no understandable reason for the evolution of this basic dualism in the cell type of organization. Each of these major parts may provide the necessary conditions for the full functioning of the other. I use the word condition here in the same sense that an aqueous medium or oxygen are indispensable conditions for the development of a frog egg, though no one would say that either is a direct cause of its development.

We think we know what the nucleus does : it provides the code or carries the information for specific macromolecular synthesis. Because of the great operability of Stentor we have found that the cortex of this cell carries the information for the number and location of oral primordia, determines which end of the anlage will coil inward to form the mouthparts, and sets the direction of asymmetry (left or right) of the organization of the band of membranelles (11). None of this important work of the cytoplasm is evident in anucleates because they cannot regenerate the complex ingestive structures from an oral primordium.

Why this failure ? Not from starvation or the cutting off of material input to the cell, because as mentioned previously maximally starved nucleated stentors are still capable of oral regeneration. Not from time limitations because regeneration requires only 10 h and anucleates survive for nearly one week. Not from lack of energy, because anucleates exhibit normal maintenance and vigorous swimming for at least 48 h. Not from absence, as such, of the nucleus, because the nucleus we know acts indirectly through what it contributes to the cytoplasm, as T.H.Morgan long ago postulated for Stentor (5). Not because an enucleate cannot produce 25,000 membranellar cilia from pre-existing basal bodies.

The simplest assumption is that the contribution of the nucleus, and in our example its support of oral primordium formation and development, is short-lived and must be continually renewed. To this prompt exhaustion of substances which have the nucleus as their source, the plant cell

Acetabularia stands in amazing contrast. What can we learn by the comparison, or by simply timing lag effects from e-nucleation in amebas or stentors in regard to RNA, protein synthesis, and specific enzymes?

The dilemna of anucleate analysis, whereby separation of nucleus and cytoplasm reveals the potentialities of neither, nor much of their interactions may possibly be circumvented by approaching the anucleate condition, asymptotically, as a mathematician might say. Already in 1909 Popoff (6) suggested that single stentors with greatly reduced nucleus are sickly and in many respects like anucleates ; Schwartz (8) asserted that their entire metabolism is disturbed. Since we can now graft stentor cells, and because the macronucleus of S.coeruleus is segmented, yet each node functionally capable of sustaining to cloning a fragment in which it is left (9), we can produce systems in which the nucleus is enormously reduced in proportion to the cytoplasm.

Previously I greatly reduced the nucleocytoplasmic ratio by grafting a cell left with only one of the ca. 15 nodes to another entirely enucleated : twice the normal cytoplasm with 1/30th the normal nuclear complement. In some but not all of these specimens, the single node became a ghost nucleus, losing its stainability with acetocarmine and eventually disappearing (12). Recently I have been producing masses of 3 to 8 stentors with but a single node left in one component. Largest cells are fused, but a third (though not more) of the cytoplasm of each is lost in the enucleations. Cytoplasmic volume was therefore somewhat less than 3 to 8 times that of a single stentor. The nuclear complement varied from 1/45th to 1/120th of the normal proportions or values of that order.

Is the single nuclear node remaining in these masses not stimulated to great enlargement and massive DNA synthesis ? I found that in single stentors reduced to 1 or 2 nodes the nodes did increase in size and a student determined that this was because of DNA synthesis (4). The present preparations therefore are no doubt "trying" to compensate for the extreme reduction of the nucleocytoplasmic ratio ; but it is obvious that they are in a critical condition and seldom capable of doing much in this direction, as indicated by the results.

These super-hypo-nucleates performed better than anucleate singlets or masses. They lived about twice as long, retained their pigmentation longer, could continue to feed after 2 days, and showed many more brown food vacuoles indicative of digestion of Paramecium bursaria. Evidently their cellular balance was precarious, for they followed one of 3 alternative courses (Fig.7).

Some persisted in better than anucleate condition, then reorganized the several sets of oral structures, simultaneously increasing the number of nuclear nodes and the total nuclear volume correspondingly (Fig.7A). This is the typical course of a single stentor reduced to one nuclear node. Mostly these were the smaller masses.

Others persisted, again better than anucleates, but then began the typical course of the late stages of an anucleate : fading of pigmentation to colorless, loss of oral cilia, the cytoplasm becoming milky white throughout. Then they died and careful search showed no nuclear node or nodes. These specimens were therefore like those previously reported in which the nucleus was "swamped" by the excess cytoplasm and disintegrated.

Most interesting were masses which continued to do well and resembled fusion masses of normal, nucleate stentors in most respects. Yet in spite of their retaining a fairly transparent cytoplasm, I repeatedly looked in vain for nuclear nodes. On squashing and dissecting and staining with aceto-carmine and aceto-orcein no specific bodies --nothing resembling the normal macronucleus of S.coeruleus-- was found. My guess is that the nucleus became dispersed but still efective within the cytoplasm. If further studies confirm this supposition, we may have a new type of "anucleate" system, an eukaryote converting itself under stress into something like a prokaryote.

Acknowledgement

This work was supported by grant CA-03637 from the National Cancer Institute, U.S.Public Health Service.

References

1. Balbiani, E.G. (1889) - Recherches expérimentales sur la mérotomie des infusoires ciliés. Contribution à l'étude du role du noyau cellulaire. Récuil Zoologique Suisse 5, 1-72.
2. Comandon, J. and de Fonbrune, P. (1939) - Greffe nucléaire totale, simple, ou multiple chez une Amibe. C.R.Soc.Biol. Paris, 130,744-748.
3. Flickinger, C.J. and Coss, R.A. (1970) - The role of the nucleus in the formation and maintenance of the contractile vacuole in Amoeba proteus. Exp.Cell.Res. 62,326-330.
4. Frazier,E.A.J. (1970) - Doctoral dissertation. University of Washington, Seattle.
5. Morgan, T.H. (1901) - Regeneration of proportionate structures in Stentor. Biol.Bull., Woods Hole, 2,311-328.
6. Popoff, M. (1909) - Experimentelle Zellstudien. II Uber die Zellgrosse, ihre fixierung und Vererbung. Arch.Zellfosch. 3, 124-180.
7. Prowazek, S. (1904) - Beitrag zur Kenntniss der Regeneration und Biologie der Protozoen. Arch.Protistenk. 3,44-59.
8. Schwartz, V. (1935) - Versuche über Regeneration und Kerndimorphismus bei Stentor coeruleus Ehrgb. Arch.Protistenk. 85, 100-139.
9. Tartar, V. (1957) - Equivalence of macronuclear nodes. J.Exp.Zool. 135, 387-402.
10. Tartar, V (1960) - Reconstitution of minced Stentor coeruleus. J.Exp.Zool. 44,187-208.
11. Tartar, V, (1961) - The Biology of Stentor, 413 pp. Pergamon Press, Oxford
12. Tartar, V. (1963) - Extreme alteration of the nucleocytoplasmic ratio in Stentor coeruleus. J. Protozool.,10 445-461.
13. Tartar, V. (1968) - Regeneration in situ of membranellar cilia in Stentor coeruleus.Trans.Amer.Microsc.Soc.,87, 297-306.

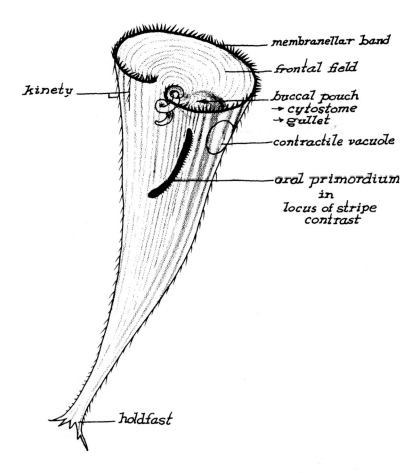

Fig. 1 - Surface structures of Stentor coeruleus.

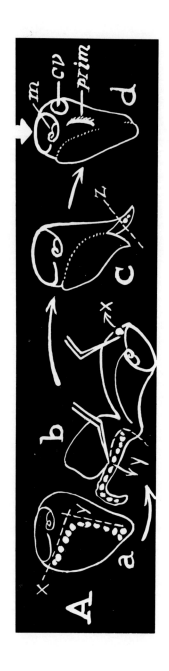
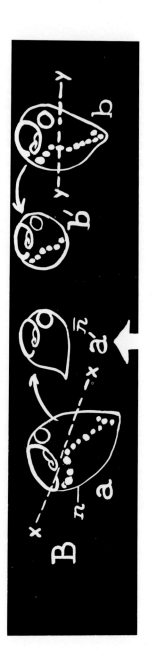

Fig. 2.A - Enucleating <u>Stentor</u> in darkfield illumination.
  a.: In the quieted ciliate two cuts (x,y) with a glass needle are made along the chain macronucleus
  b.: Chain of nuclear nodes is excised with minimum cytoplasm (y). Anterior nodes of chain are teased out if left in cell (x).
  c.: Posterior nodes also excised if partially obscured (z). Resulting anucleate (↓) retains most of cytoplasm, and important structures of mouth, contractile vacuole and primordium (if present) need not be disturbed.

Fig. 2.B. - Simple transection yielding anucleate and nucleate portions.
  a.: Cell cut along line x - x gives nucleate portion (n) and anucleate ($\bar{n}$).
  a': Anucleate piece carries ingestive structures and original contractile vacuole.
  b.: It may best be compared with nucleate pieces of equal size (b') cut by transverse section y - y of another cell.

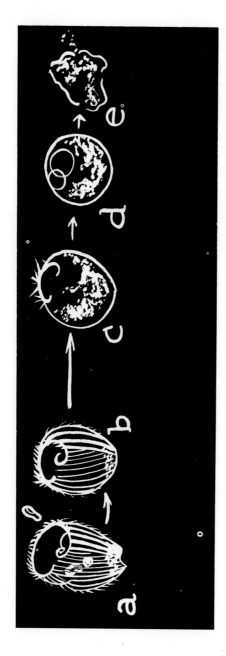

Fig. 3 – Course of survival of anucleate S.coeruleus
a.: Cell is normally active first day and can capture and partially digest its prey. Opaque carbohydrate reserves concentrated at posterior end are slowly utilized.
b.: After day 3, the cell slows and the mouthparts begin regressing.
c.: By day 4 the cytoplasm becomes milky-opaque and the surface pigmentation is gone. Body cilia are not seen and the few remaining membranellar cilia only weakly twitch.
d.: Perfect, glistening sphere and vesicles atest loss of surface structures and of osmotic regulation by day 5.
e.: By day 6 anucleates are usually dead.

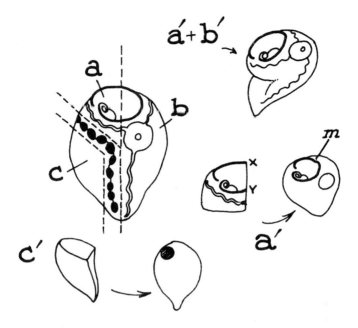

Fig. 4 - Contractile vacuole in anucleate S.coeruleus. Cell can be enucleated as shown without disturbing the contractile vacuole and its perported circumoral and posterior feeding canals. Then anucleate part a' + b' shows continued normal beating of the vacuole for 2 days.
    a' : Anucleate portions containing part of the feeding canal system can regenerate a new and functional contractile vacuole.
    c' : portions without these parts nor a part of the nucleus cannot. They produce a vesicle which does not pulsate.
Note in a' that cut ends (x,y) of the membranellar band neatly mend (m) in the anucleate, metachronal rhythm of the membranelles in the two portions then becoming again coordinated.

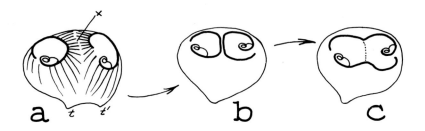

Fig. 5 - Unification by selective resorption of cortical parts in graft of two anucleate stentors.
    a.: When fused, heads are separated by body striping (x) ; tails (t - t') also separate.
    b.: Heads fuse by resorption of striping and tail poles likewise.
    c.: Apposed portions of the two membranellar bands are selectively resorbed, leaving bistomial head and continuous frontal field.

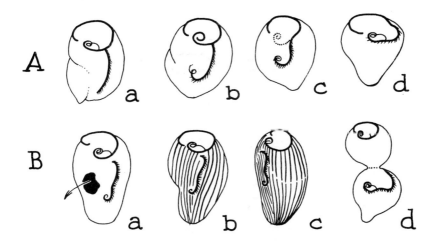

Fig. 6. - A : Completion of reorganization after enucleation.
  a. Stage 4 reorganizer with new membranellar band.
  b. Original mouthparts regress as replacing set is formed.
  c. Precedent mouthparts and adjacent membranellar band selectively resorbed as replacement of ingestive structures is completed.

B : Continuation of division after enucleation.
  a. Stage-6 pre-divider has compacted macronucleus easily excised.
  b. There is still no indication of a division line.
  c. Fission line formation by severing of kineties and pigment bands around the circumference of the cell.
  d. Full constriction along fission line results in two anucleate daughter cells.

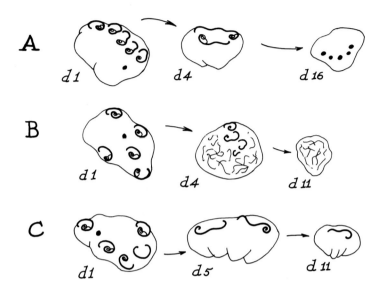

Fig. 7 - Fate of super-hypo-nucleates of S.coeruleus

    A : Fusion mass of 5 stentors with one small macronuclear node. By day 4 the specimen had reorganized as a doublet. Reducing in size as it used its substance, the object died on day 16 and showed 5 nuclear nodes, about normally proportionate to the cytoplasmic volume.

    B : Five stentors, one node. By day 4 the pigment was gone and cytoplasm became milky-opaque as in anucleates. Surviving until day 11, the specimen then showed no nucleus.

    C : Five stentors with one node. By day 5 the healthy mass had reorganized to a two-headed stentor and had a polarized orientation with multiple tail-poles. On day 11 the specimen, though it had ingested and digested some food organisms was reduced in size to a normal singlet. Careful examination showed no discrete objects resembling the normal nuclear nodes.

CYTOPLASMIC DAMAGE LEADING TO DELAY OF ORAL REGENERATION IN STENTOR COERULEUS

Brower R. Burchill

Department of Physiology and Cell Biology
The University of Kansas
Lawrence, Kansas 66044, U.S.A.

Abstract

Ultraviolet light (UV)-irradiation leads to a rapid but transient interruption of oral regeneration, the replacement of missing feeding organelles, in the ciliate Stentor coeruleus, resulting subsequently in a delay in the completion of development. Since the large size of this cell renders it unlikely that significant amounts of UV can penetrate the cytoplasm to the macronucleus the primary site of UV absorption probably resides in the cytoplasm. UV-microbeam experiments indicate, however, that damage to the macronucleus, or something in the immediate vicinity of this organelle, is involved in the radiation response. Interference with some macronuclear or other process essential to regeneration is proposed to result from the action of a diffusible cytoplasmic "phototoxin". A recovery process, which either acts to rid the cell of the "toxin" or to repair some secondary damage, also occurs. The UV absorbing moiety in the cytoplasm has not been identified, nor has the "toxin", nor the process (es) with which it interfers. Experiments designed to help answer these questions are discussed.

Paper read by title

Introduction

My intent in this paper is not to review either the field of cytoplasmic radiation sensitivity or that of protozoan radiation sensitivity. Rather it is to review certain aspects of my own work with the ciliated protozoan Stentor coeruleus which have already been published and to present some new information, and to discuss the implication of the data to the subject of ultraviolet light (UV)-sensitivity in this specific cell. Also, for the sake of brevity, I will not repeat here that data which has been published before, but will refer to the appropriate literature if the reader wishes more in-depth information.

Materials and Methods

Organism

Like other ciliates certain regions of Stentor coeruleus are organized into highly specialized feeding organelles, including a band of ciliary organelles encircling the anterior end of the conically shaped cell, and a gullet[1]. The work to be described concerns the effects of UV on the regenerative replacement of missing feeding organelles. This replacement, known as oral regeneration, requires about 8 hours for completion and proceeds through a sequence of cytologically distinct stages which, due to the large size of the cell (spherical diameter about 0.25 mm), can be easily identified using the dissecting microscope [1,2]. The sequence of identifiable stages acts as a series of fiducial marks by which the effects of experimental treatment (eg. irradiation, metabolic inhibitors) can be immediately and precisely ascertained. Development in treated cells can be compared to that in untreated controls from the same culture. It is this capacity of rapid determination of effect which makes oral regeneration in Stentor such a useful tool for the study of the primary effects of extraneous agents, including radiations, on differentiation in eucaryotic cells.

The most readily apparent events which occur during, and those by which effects on development can be most easily determined, include a vast proliferation of new cilia (and their supporting basal bodies) to form the new band of ciliary organelles, and extensive shape changes in the metabolic nucleus of the cell, the macronucleus (MN). The new ciliary organelles arise in a specific region of the cell's cortex and then migrate to the site previously vacated by the missing feeding organelles. The new gullet forms at the posterior end of the ciliary band and migrates with it. The MN normally extends the length of the cell as a series of small, interconnected nodes. Late in regeneration (stage 6), however, it condenses into from one to three large spheres and then renodulates to complete development coincident with the final movements of the new ciliary organelles (1).

Large numbers of stentors were made to undergo simultaneous oral regeneration by removing their existing feeding organelles at the same time. This was accomplished by a brief exposure to a 15 % solution of sucrose, followed by thorough washing (cf.3 for a discussion of the sucrose technique and of the synchrony of development in groups of sucrosed stentors). From this population samples were selected both for experimental manipulation and as untreated controls. The progress of development was determined by frequent (1/2 to 1 hour) observations of each cell with a Zeiss Stereomicroscope III.

Other
Whole cell and partial cell UV-irradiations were performed, respectively, with germicidal lamps and a reflex camera microbeam as previously described (3,4). An autoradiographic study was performed using L-leucine-4,5-^{3}H (50 μc /ml ; 5 c /mM ; New England Nuclear) and Kodak NTB-2 liquid emulsion (see 2 for detailed procedure). Actinomycin D was a gift of Merck, Sharp and Dohme. Puromycin and cycloheximide were purchased, respectively, from Nutritional Biochemical Company and The Upjohn Company.

Results and Discussion

General
The predominant response of regenerating stentors to UV-irradiation was a dose-dependent, transient suspension of development (3). This suspension usually began in the

stage of development immediately succeeding the one in which irradiation was performed. After this transient interruption the remainder of development was completed at a normal rate. The overall result was a delay in the completion of regeneration essentially equivalent to that period of interrupted progress shortly after irradiation. The term UV-sensitivity, as used in this paper, will refer to the extent of this delay (see 3 & 5 for more complete discussions of radiation-induced delay of cellular processes in Stentor and other cells).

Hypothesis

The working hypothesis of the basis of UV-induced regeneration delay in S.coeruleus which has been formulated from the evidence to be discussed is as follows :

> Absorption of UV energy by the cytoplasm of Stentor leads to the formation of a "phototoxin" which subsequently interfers with some process(es) in the cell essential to development ; this interference leads to a temporary cessation of regeneration. The cessation is temporary because of a recovery process which reverses the radiation effect. The UV absorbing moiety in the cytoplasm has not been identified, nor has the "toxin", nor the process(es) with which it interfers.

UV-Microbeam

In an attempt to determine if any specific region of regenerating stentors was highly sensitive to UV various sites were selectively irradiated with comparable incident doses of radiation from a 100 micron diameter spot of 265 nm UV (4). No attempt was made to measure the intensity of the microbeam. Regions including only cytoplasm or cytoplasm plus macronuclear nodes were irradiated in cells held immobile and flattened to about 115 microns beneath a quartz coverglass. Relative sensitivity of the various regions was determined by comparing the percentage of cells significantly delayed by irradiation of the various sites during the early stages of development (whole cells are most UV-sensitive during early stages : 3 ; see 4 for a complete discussion of techniques and results).

The results indicated that cells were most frequently affected (89 % delayed) if the target area included 2 or 3 macronuclear nodes positioned midway between the anterior and posterior ends of the cell. Because of the large size of the macronuclear nodes no more than 2 or 3 could be included in the beam at any one time. The next most sensitive regions, in order of decreasing sensitivity, were : macronuclear nodes near the extremes of the cell (59 % delayed) ; cytoplasm in the central region of the cell (50 % delayed) ; and cytoplasm near the ends of the cell (36 % delayed).

These results suggest two things. First, the MN or something in its immediate vicinity is the most UV-sensitive site in the cell. Second, the cytoplasm is also sensitive. The first of these conclusions must be tempered, however, since it has been estimated (6) that 90 % of the UV energy incident on a typical cell has been attenuated by nucleic acids in the cytoplasm at a depth of 10 microns. If <u>Stentor</u> is a typical cell (and it certainly does have an abundance of mitochondria, presumably containing DNA, in the sub-pellicular region ; 7) it is unlikely that significant UV could penetrate to the region of the MN, either in whole or compressed cells. In both cases this organelle lies further than 10 microns beneath the cell surface. Experiments are currently in progress to determine precisely how much direct damage does occur in the MN of irradiated stentors by measuring the formation of pyrimidine dimers in macronuclear DNA. For the present it will be assumed that the MN is <u>not</u> a significant, primary UV-sensitive site. The primary site of energy deposition, therefore, must reside in the cytoplasm.

<u>Phototoxin</u>

The higher percentage of delay which occurred when macronuclear nodes were in the microbeam would seem to suggest that the MN, if not a primary UV-sensitive site, may be damaged secondarily as a result of UV absorption by the cytoplasm. That interference with the MN might lead to delay of regeneration is not surprising. This organelle is absolutely essential to normal development in early stages (prior to stage 5) (1). However, it has also been shown that as few as 6 macronuclear nodes (out of a normal number of 12 to 18) are sufficient to guide oral regeneration to a successful completion at a normal site. With fewer than 6 nodes the process is delayed. This elegant study was performed by Vance Tartar (8) who made use of the extreme amenability of S.coeruleus to microsurgical manipulation and removed different numbers of macronuclear nodes at various times during oral regeneration. The point of significance for the present study is that if the macronucleus is a site of secondary UV damage more nodes than those in the immediate vicinity of the microbeam must be affected. This suggests the action of a "phototoxin",generated as a result of the initial absorption of UV energy by the cytoplasm. That cells were more often delayed when irradiated with the microbeam in the central region of the cell is compatible with this idea, for the "toxin" might be expected to interact more readily and quickly with the macronuclear nodes starting from a central location than from one end of the macronuclear chain or the other.

There have been many reports of cellular processes altered by cytoplasmic UV-irradiation (for review see 4, 6) and in those instances where damage resulted in parts of the cell not directly in the UV beam the existence of a "phototoxin" has been proposed.

Dose-Response

A dose-response study was performed of whole, regenerating cells irradiated with germicidal lamps soon after sucrose treatment (the induction of regeneration). The increase in delay with dose was approximately linear between doses of 3,500 and 25,000 ergs/mm^2 (analysis of variance indicated $P < 0.005$; 3).

Since regeneration is only delayed and not permanently suspended by these doses of UV, and since the amount of delay increases with dose, it may be that some UV-sensitive site in the cell is capable of varying degrees of partial damage. Bacq & Alexander (9) have suggested that when a very large number of radiation sensitive sites are present in a cell damage to less than 25 % should lead to a linear dose-response. Higher doses would lead to an exponential response. Unfortunately, higher doses than those indicated killed the stentors, perhaps by a mechanism quite different from that leading to delay, precluding determination of a slope at higher doses (see 3 for a more complete discussion, including other possible explanations for linear dose-response). Some entity present in abundance in the cytoplasm could easily satisfy the criterion suggested by Bacq & Alexander of a site capable of varying degrees of partial damage thus leading to a linear dose-response.

Recovery
Dark repair is also a factor in UV-induced regeneration delay.

Recovery in non-regenerating cells.
Cells irradiated immediately prior to the induction of regeneration (sucrose treatment) suffer a regeneration delay identical to that of cells irradiated immediately after sucrose treatment. This indicates that the UV absorbing entity is present even in non-regenerating cells.

Incubation at room temperature.
As shown in Fig.1, cells irradiated immediately before sucrose treatment (recovery time 0) were delayed an average of 2.4 h. As the interval (recovery time) between irradiation and sucrose treatment was increased, the amount of regeneration delay was reduced, indicating that recovery was occuring. Little or no delay remained after 4 h. An analysis of variance test indicated that the reduction in delay occurred at an approximately linear rate with time ($P < 0.01$). The variability in response indicated by the large standard deviations in Fig.1 results from the fact that the radiation sensitivity of Stentor varies considerably between cells from different cultures and from day to day (3), and the fact that the graph is a composite of data from 8 experiments performed on different days and using cells from several stock cultures.

To investigate the metabolic requirements of the repair process irradiated cells were exposed to cold or to a reversible inhibitor of macromolecular (protein and perhaps DNA) synthesis prior to sucrose treatment.

Incubation at 4°C
As shown in Fig.2, there was no significant recovery when irradiated cells were incubated for various intervals at 4°C prior to sucrose treatment. This suggests that some metabolic event or some temperature sensitive structural change is essential to recovery. When unirradiated cells were exposed to the cold for equivalent periods and then returned to room temperature at the start of regeneration they developed at the same rate as uncooled control cells.

Incubation in cycloheximide

To determine if the recovery process required macromolecular syntheses cells were briefly exposed to a reversible inhibitor, cycloheximide, after irradiation. This chemical inhibits protein synthesis in Stentor and other cells (2 ; Table 1), and interfers directly or indirectly with DNA synthesis in some cells (10 , 11) perhaps including Stentor (preliminary results of Burchill suggest a reduction in tritiated thymidine incorporation in non-regenerating stentors by 5 µg/ml cycloheximide).

The concentration of cycloheximide used in the recovery experiments was selected because it was sufficient to block the development of stentors exposed soon after the initiation of regeneration (2) and because higher concentrations were not as readily reversible. Table 1 shows that at a concentration of 5 µg/ml cycloheximide reduced tritiated leucine incorporation into acid insoluble material of non-regenerating cells by 34 % during a 1-h incubation period. Concentrations up to 50 fold higher reduced incorporation further but did not completely stop it. This is consistent with a prior study in Stentor which found that cycloheximide did not completely eliminate leucine incorporation in regenerating cells, even though the concentrations tested effectively blocked regeneration (2). There is evidence in the literature to suggest that amino acid incorporation which continues in cells in the presence of cycloheximide does not represent the synthesis of functional protein synthesized on cytoplasmic ribosomes (12 , 13).

The effect of cycloheximide on recovery is reported in Table 2. The results are partitioned into two sections, showing experiments which involved post-irradiation exposures to cycloheximide of both 1 and 2 h duration.

The delay arising from treatment with cycloheximide (column 2) is unexplained, and speculation does not seem warrented. However, the data conclusively indicate that, regardless of the reason for the cycloheximide alone effect, UV recovery did not occur normally in the presence of the inhibitor. Had it done so the delays of cells treated with both agents should not have been greater than those caused by cycloheximide alone. In other words, if recovery continued at a normal rate in the presence of cy-

cloheximide, repair should have been nearly complete (only
0.7 h remaining ; see column 3) at the end of an incubation
period of 2 h. After washing at the time regeneration was
induced repair would soon have been completed, with the result that the only delay remaining would be that caused by
the cycloheximide treatment alone (1.9 h ; see column 2).
Instead, the delay was 3.4 h, indicating that the effect
of the two agents were not totally independent of each
other.

Late stage recovery

During the first half of regeneration, up to stage 4,
stentors are very sensitive to UV (3). During and after
stage 4 sensitivity is much lower. In addition cells exposed to chemical inhibitors of RNA and protein synthesis
(actinomycin D, puromycin, cycloheximide, emetine) in stages later than stage 4 are not prevented from completing
regeneration. These chemicals do permanently suspend development if applied at earlier stages (3 ; Burchill, unpublished observations). The combination of these situations,
the continued slight sensitivity to UV but relative insensitivity to certain metabolic inhibitors in late stages,
presented the possibility of testing the effects of interference with macromolecular syntheses on post-irradiation
development of cells irradiated in stage 5. The results
are presented in Table 3.

When cells irradiated in stage 5 were immediately exposed to either actinomycin D or puromycin the completion
of development was either prevented or greatly delayed.
This suggests, therefore, that RNA and protein synthesis
must occur if the cell is to recovery normally from the UV
damage leading to delay (see 2 for effect of actinomycin D
and puromycin on incorporation of uridine and leucine in
regenerating Stentor).

Nature of UV-sensitive sites and events

The site(s) of cellular injury which might necessitate repair and which are compatible with the available data are many and further investigation of the dark repair
process may aid in their identification.
1) Repair might simply encompass elimination of the "toxin". Only when this is accomplished might the process(es)
essential to regeneration with which it interfers resume.

The time required to remove the "toxin" (regeneration delay) could relate to the amount of "toxin" generated and thus to the UV dose. Protein synthesis required for recovery might include the synthesis of an enzyme needed either to inactivate the "toxin" or effect its transport from the cell.

2) Recovery might alternatively involve the repair of some secondary damage. One observation suggests that at least one secondary damage is to a process which is of a very basic nature to cell differentiation in Stentor. Regardless of the time during regeneration of UV-irradiation, development was quickly halted ; eg. the sensitive process was not unique to any particular stage of regeneration. Variation in the degree of UV-sensitivity during regeneration (3) might reflect either altered amounts of "toxin" formed or changes in the physical or metabolic state of the essential process at various times. There are several events known to be essential throughout most if not all of oral regeneration in Stentor, and their possible involvement in UV-induced regeneration delay is being further investigated.

a) One such event involves the microtubular components in this ciliate protozoa. Agents which interfer with microtubules, colchicine, colcemid, vinblastine sulfate, etc. (15 ; 16 ; Burchill, unpublished observation) suspend development in Stentor, even when administered as late as stage 5. As in UV-irradiated cells, development continues for a short time before being halted. Not only are microtubules apparently essential throughout development, thereby fulfilling the requirement of an event which is basic to all of regeneration, but they have been implicated as a mediator in UV-induced spindle disappearance in microbeam-irradiated Xenopus cells (17 ; 18).

In addition, both Stentor and Tetrahymena (another ciliate protozoan) can recover from the influence of certain of these anti-microtubule agents (15 ; 19 ; Burchill, unpublished observation). It is conceivable that the capacity for such recovery is a basic capability of these ciliates. Study of the mechanism of recovery from the chemical agents may provide information about the nature of the UV effect. An investigation is currently underway in this laboratory to determine whether macromolecular syntheses are required for the recovery from colchicine toxicity, as they are from UV-induced regeneration delay.

b) Another process whose secondary damage by the "toxin" might necessitate a repair process is RNA synthesis. Since cycloheximide has been found to interfer with DNA synthesis, perhaps as a consequence of its inhibition of protein synthesis, in some cells (10 ; 11) it may be that the interference of cycloheximide with recovery in UV-irradiated Stentor results from its effect on a DNA repair mechanism. Certainly macronuclear DNA is a prime candidate for such a secondary sensitive site early in regeneration and such damage could alter RNA synthesis (2). As in UV-irradiated cells regenerating Stentor exposed to concentrations of actinomycin D which completely suppress macronuclear RNA synthesis cease development only after a short period of time. Radiation sensitivity after stage 4 must, of course, be of a different nature since RNA synthesis is no longer essential (2).

c) Some event linked to mitochondrial DNA (mDNA) may also be a candidate for a secondary sensitive event. Damage to mDNA, for example, might interfer with the synthesis or availability of ATP. ATP synthesis is essential throughout most of regeneration (Rahner & Burchill, unpublished observation). In addition, mDNA may be intimately related to cellular metabolism in ways that we are completely unaware of at this time. Experiments are currently underway in this laboratory to assay the DNA of several cellular compartments in UV-irradiated Stentor (MN, mitochondria, basal bodies?) for pyrimidine dimers. While such dimers are not likely to result from secondary "toxin" damage the finding of dimers in mDNA, for example, might suggest this organelle as a primary UV-sensitive cytoplasmic site. If dimers are found in any cellular DNA an attempt will be made to quantitatively relate the degree of such damage to the period of delay in comparably irradiated cells, and the repair of that damage to the kinetics of the recovery process.

In addition, an attempt will be made to photoreactivate both the dimers and UV-induced regeneration delay. Photoreactivation of cytoplasmic damage leading to UV-induced division delay in Amoeba proteus has been found (6), as has photoreactivation of pyrimidine dimers in mDNA (20).

Preliminary attempts to photoreactivate UV-induced regeneration delay in Stentor have been unsuccessful, but a more determined effort will be made, especially if damage to some DNA compartment is implicated in other ways.

Conclusions

Ultraviolet light interfers with various processes in a large number of different cells through initial damage to the cytoplasm (4-6). At the present time no UV-sensitive cytoplasmic site has been positively identified, nor has the nature of any "toxin" leading to any specific secondary effect. One goal of research in this laboratory is to characterize in specific terms the nature of the cytoplasmic damage responsible for UV-induced regeneration delay in Stentor coeruleus.

Acknowledgements

Parts of this work have been supported by the following grants and agencies : U.S. Atomic Energy Commission contract W-31-109-ENG-78 to the Division of Radiation Biology, Case Western Reserve University ; U.S. Atomic Energy Commission at the Los Alamos Scientific Laboratory; a grant from the Research Corporation, Minneapolis, Minnesota ; Biomedical Sciences Support Grant RR07037 ; and a General Research Grant, # 3041-5038, from the University of Kansas.

References

1. V.Tartar, The Biology of Stentor, Pergamon Press, New York (1961).
2. B.Burchill, J.Exp.Zool., 167, 427 (1968).
3. B.Burchill, J.Exp.Zool., 169, 471 (1968).
4. B.Burchill & R.Rustad, J.Protozool., 16, 303 (1969).
5. R.Rustad, Photochem.Photobiol., 3, 529 (1964).
6. J.Jagger, D.Prescott & M.Gaulden, Exp.Cell Res., 58, 35 (1969).
7. J.Paulin & J.Bussey, J.Protozool., 18, 201 (1971).
8. V.Tartar, J.Protozool., 10, 445 (1963).
9. Z.Bacq & P.Alexander, Fundamentals of Radiobiology, Pergamon Press, New York (1961).
10. H.Emis & M.Lubin, Science, 146, 1474 (1964).
11. F.Wanka & J.Moors, Biochem.Biophys.Res.Comm., 41, 85 (1970).
12. S.Rajalakshmi, H.Liang, D.Sarma, R.Kisilevsky & E.Farber, Biochem.Biophys.Res.Comm., 42, 259 (1971).
13. B.Hogan & P.Gross, J.Cell Biol., 49, 692 (1971).
14. B.Burchill, Radiat.Res., 27, 543 (1966).
15. J.Haight,Jr. & B.Burchill, J.Protozool., 17, 139 (1970).
16. E.Makrides, S.Banerjee, L.Handler & L.Margulis, J.Protozool., 14, 548 (1970).
17. D.Brown & R.Zirkle, Photochem.Photobiol., 6, 817 (1967).
18. D.Heaton & D.Brown, Abstracts of Papers for the Nineteenth Annual Meeting of the Radiation Research Society (Abs. Dc-9) (1971).
19. F.Wunderlich & V.Speth, Protoplasma, 70, 139 (1970).
20. J.Cook & T.Worthy, Biophysical Society Abstracts - Fifteenth Annual Meeting (Abs. WPM-K4) (1971).

Table 1 - Effects of Cycloheximide on leucine - ^{3}H Incorporation in Non-Regenerating Stentor (Unpublished data of B. Burchill)

	Concentration (µg/ml)	90°C TCA (15 min)
Control :		
Macronucleus	-	55 ± 15.8 [1]
Cytoplasm	-	57 ± 18.6
+ Cycloheximide :		
Macronucleus	5	29 ± 11.0
Cytoplasm		38 ± 8.5 (34%) [2]
Macronucleus	50	25 ± 1.2
Cytoplasm		27 ± 8.7 (53%)
Macronucleus	250	5 ± 1.5
Cytoplasm		14 ± 5.5 (76%)

(1) Grain counts/unit area (above background) and standard deviations of numbers of grains in sections counted.

(2) Percentages show extent of reduction by cycloheximide of cytoplasmic leucine incorporation.

Table 2 — Effect of cycloheximide on recovery from UV-induced regeneration delay in Stentor coeruleus (Unpublished data of B. Burchill)

	Column 1 (a) Ultraviolet irradiation	Column 2 (b) Pre-regeneration cycloheximide treatment	Column 3 (c) Post-irradiation incubation in culture medium	Column 4 (d) Post-irradiation incubation in cycloheximide
1-Hr incubation period prior to regeneration				
Average delay of regeneration (h)	2.0 ± 0.79	2.2 ± 0.77	1.4 ± 0.61	3.0 ± 0.50
Number of cells employed	123	35	17	16
2-Hr incubation period prior to regeneration				
Average delay of regeneration (h)	—	1.9 ± 0.64	0.7 ± 0.44	3.4 ± 0.48
Number of cells employed	—	131	88	99

(a) Cells induced to regenerate immediately after irradiation;

(b) Cells exposed to cycloheximide for 1 or 2 h before regeneration;

(c) Cells maintained in normal culture medium for 1 or 2 h after irradiation and before regeneration;

(d) Cells exposed to cycloheximide for 1 or 2 h after irradiation and before regeneration.

Table 3 - Influence of post-UV treatment with actinomycin D and puromycin on regenerating <u>Stentors</u> irradiated in stage 5 (Unpublished data of B.Burchill; cf. 14)

Treatment	Effect			Cell Total
	Regeneration Blocked	Regeneration Delayed	Regeneration not Affected	
Actinomycin D (2×10^{-4} M)	0[a]	26	34	60
Puromycin (2.3×10^{-4} M)	0	24	8	32
UV	4	42	4	50
UV + Actinomycin D	15	49(20)[b]	1	65
UV + Puromycin	25	6 (4)	0	31

(a) Numbers of cells affected

(b) Number of cells delayed significantly more than cells treated with UV alone.

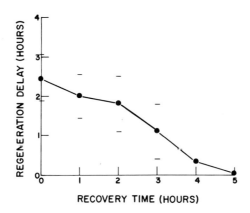

Fig. 1. Recovery at room temperature (22°C) after UV-irradiation. The points represent the mean delays in regeneration of cells which were irradiated with an average dose of 11,700 ergs/mm^2 and then maintained from 0-5 hr at 22°C prior to induction of regeneration. The graph includes data from 8 experiments, and each point represents an average of 40 cells. The limit bars indicate the standard deviations from the mean (Unpublished data of B.Burchill).

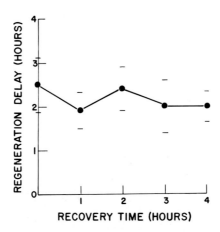

Fig. 2. Recovery at 4°C after UV-irradiation. The points represent the mean delays in regeneration of cells which were irradiated with an average dose of 12,300 ergs/mm^2 and then maintained from 0-4 hr at 4°C prior to induction of regeneration. The graph includes data from 5 experiments, and each point represents an average of 27 cells. The limit bars indicate the standard deviations from the mean (Unpublished data of B.Burchill).

JUNE 22

SESSION 4
Chairman: Lucien Vakaet

THE REGULATION OF PROTEIN SYNTHESIS IN ANUCLEATE FROG OOCYTES

R. E. Ecker

Division of Biological and Medical Research
Argonne National Laboratory
Argonne, Illinois 60439, U.S.A.

Introduction

When the amphibian oocyte is full grown and ready to be ovulated, it is about 1-2 mm in diameter and has a dry mass of several milligrams. This size represents almost a millionfold increase from that of its primary progenitor and requires up to three years of growth. Much of the increase in size is due to the accumulation of yolk, most of which is probably synthesized outside the oocyte and transferred in during oogenesis (Wallace and Jared, 1969). But the amphibian oocyte itself is active in transcription during oogenesis and, particularly during the "lampbrush stage," synthesizes and accumulates both ribosomal and informational RNA (see Davidson, 1968). Protein synthesis also occurs within the oocyte during oogenesis (Davidson, 1968). Although there is little information concerning the specific nature or possible localizations of these proteins, it is likely that most of them are involved in the structure and sub-structure of the full-grown oocyte and in its unique enzymatic character.

Throughout oogenesis in the amphibian, the oocyte nucleus remains in prophase of the first meiotic division and contains twice the normal diploid amount of DNA. In Rana pipiens, when the oocyte is full grown, the nucleus (germinal vesicle) constitutes about 1% of the total oocyte volume and contains, in addition to the DNA (both nuclear and nucleolar; see Dawid, 1965; Brown and Dawid, 1968), about 5 µg of protein and about 0.1 µg of RNA. The chromosomes have condensed from the earlier "lampbrush" configuration and are surrounded by a multitude of nucleoli.

The oocyte remains suspended in this more-or-less inactive state until induced to resume maturation by hormonal stimulation. The resulting hormone-induced maturation requires about 30-36 hrs at 18°C and takes the oocyte through germinal vesicle breakdown, the remainder of first meiosis and second meiosis up to metaphase. At this point maturation again ceases and will resume only following activation.

Although normally in vivo the events of hormone stimulated maturation occur concomitantly with ovulation and the migration of the ovulated oocytes, through the body cavity and oviducts, into the uteri, these events can be induced in vitro in oocytes removed manually from their ovarian follicles. And, following hormone-induced maturation in vitro, these oocytes are capable of being activated and of undergoing normal development (Smith, Ecker, and Subtelny, 1968).

Measurements of protein synthesis

We measure protein synthesis in R. pipiens oocytes and early embryos by following the incorporation kinetics of micro-injected tritiated amino acids. Measured injections of nanoliter volumes are made by means of calibrated micropipets and a Leitz micromanipulator (Ecker and Smith, 1968). An example of such a pipet is shown in Figure 1. Oocytes are injected individually with tritiated amino acid and, at various times thereafter, are killed in groups of 2 or 3 by transfer to hot (60°C) 0.5 N perchloric acid (PCA). After extraction for at least 30 min in hot PCA, the oocytes are washed free of PCA by three passages through distilled water, transferred to one inch diameter filter paper discs, dried and combusted to HTO (tritiated water) in a Packard Model 305 Tri-Carb Sample Oxidizer (Packard Instruments, Downers Grove, Illinois.) The HTO is collected in scintillation fluid consisting of 100 g naphthalene; 5 g 2,5-diphenyloxazole (PPO); 0.3 g 1,4-bis [2-(5-phenyloxazolyl)]-benzene (POPOP); 720 ml dioxane; 135 ml toluene; and 45 ml absolute methanol, and counted in a Beckman CPM 100 liquid scintillation spectrometer (Beckman Instruments, Palo Alto, California).

Typical kinetics, showing cpm incorporated as a function of time after injection of tritiated leucine, are shown in Figure 2. It has been shown that these curves can be described by the equation

$$P^* = \alpha(1-e^{-\beta t}), \qquad (1)$$

where P^* is cpm incorporated into protein after any time, t; $\alpha = kL_o$, where L_o is the amount of activity initially injected and k is the recovery and counting efficiency; $\beta = f/V$, where f is the rate of flow of leucine through the amino acid pool and V is the size of that pool. Equation (1) is derived from the assumption of a steady-state amino acid pool into which the labelled amino acid is injected and with which it rapidly equilibrates (see Ecker and Smith, 1966; 1968). Equation (1) can be fitted by computer to the experimental data by use of a non-linear, least squares curve-fitting program; thus, values of the two parameters α and β, can be obtained for each experiment.

The parameter β (or f/V) is of particular importance, as f, the rate of flow of leucine through the pool, is directly related to protein synthetic rate. Thus, if the value of V is known or can be determined, these kinds of kinetic experiments can be used to estimate rates of protein synthesis in amphibian oocytes. Such estimates have been made (Ecker and Smith, 1966; 1968), but they have been based on chemical measurements of extracted amino acid pools. If such extracts contain substantial amounts of amino acids that <u>in situ</u> are not active in protein synthesis, estimates of f based on these measurements could be substantially in error.

For this reason a technique has been developed to measure both f and V from amino acid incorporation kinetics. This technique takes advantage of the observation (Ecker and Smith, 1968) that increase in the endogenous amino acid pool of the oocyte by injection does not appear to alter the rate of flow of amino acid through the pool. Only the uptake kinetics are changed and these changes can be predicted using equation (1), with an appropriately adjusted value of V.

Thus, if one looks at the kinetics of incorporation of two different concentrations of injected leucine, two different incorporation curves are obtained as indicated in Figure 2. The differences between the two curves are due to differences in the amount by which the endogenous leucine pool was expanded by the two isotope preparations. These differences can be described mathematically as follows:

$$\beta_1 = \frac{f}{V + V_1} \text{ and} \qquad (2)$$

$$\beta_2 = \frac{f}{V + V_2}, \qquad (3)$$

where β_1 and β_2 are the incorporation rate constants obtained by computer fit of equation (1) to the two sets of experimental data, and V_1 and V_2 are the amounts of exogenous amino acid injected in each case. These latter values are obtained from the known concentrations and volumes of the amino acid solutions injected. In the experiment shown in Figure 2, the value of V_1 (corresponding to the curve of closed circles) is 0.1 μmoles/ml times 12.5×10^{-6} ml or 1.25×10^{-6} μmoles/oocyte; and by the same kind of computation V_2 (corresponding to the open circles) is 2.88×10^{-6} μmoles/oocyte.

The value of V and f can be determined from the simultaneous solution of equations (2) and (3). When this is done for the example in Figure 2, these values are : $V = 9.2 \times 10^{-7}$ μmoles/oocyte and $f = 2.2 \times 10^{-7}$ μmoles/oocyte/min. If one assumes that the labelled proteins are about 10% leucine (see Ecker and Smith, 1966), the indicated rate of protein synthesis would be about 0.02 μg/oocyte/hr.

It will be noted that this rate of protein synthesis for oocytes at about first meiotic metaphase is lower by a factor of 30 than that reported previously (Ecker and Smith, 1968). From the same reference it will be noted that the size of the leucine pool listed above is 80 times smaller than the lowest value obtained from chemical determinations of the extracted pools. Thus the amount of free amino acids active in oocyte protein synthesis seems to be very much smaller than the total amount extractable from the oocyte. We have no idea at this time concerning the possible localization of this large body of amino acid. Obviously the active pool is freely accessible in the cytoplasm; Figure 2 shows that it can be easily and predictably expanded by injection of amino acid directly into the cytoplasm.

This great discrepancy, then, between the extractable pool and that apparently active in protein synthesis, accounts for the large difference in the rate of protein synthesis between that reported here and that we reported earlier (Ecker and Smith, 1968). Therefore, we find that this technique not only provides a somewhat more convenient way of measuring protein synthesis in amphibian oocytes, but it also probably offers the only way of making these determinations with any accuracy at all.

Enucleation and its developmental consequences

When a full-grown oocyte has been removed from its ovarian follicle, the germinal vesicle can be manually removed with little difficulty. The enucleation process is outlined in Figure 3. The small incision in the oocyte, made to remove the nucleus, heals over quickly and enucleated oocytes are ready for further manipulation within an hour or two.

If oocytes are enucleated and then exposed to progesterone, they resume maturation and undergo all of the maturational events observed in nucleated controls, except those related to the chromosomes (meiosis). And, after the usual 30-36 hr maturation period, these enucleated oocytes are capable of being activated, and exhibit the same cytoplasmic criteria of successful activation as the nucleated controls (see Smith and Ecker, 1969). Obviously, then, neither the intact germinal vesicle nor its contents are required to allow the normal, hormone-induced maturation process up to the point of activation; the target(s) of the steroid hormone within the oocyte is cytoplasmic, as is the machinery for driving and regulating the steps in the maturational process.

However, development beyond this point is not possible in oocytes enucleated by removal of the germinal vesicle. Transfer of blastula nuclei into nucleated oocytes, which were induced to mature in vitro and then activated, gives a high percentage of normal development. Yet, oocytes that are enucleated and then caused to mature in vitro give no normal cleavage when they are activated and then receive a transplanted blastula nucleus. This loss of developmental potential can be restored simply by injecting the enucleated oocyte, prior to nuclear transfer, with material from the germinal vesicle of a non-hormone-treated oocyte (Smith and Ecker, 1969).

Thus, it is apparent that, although the germinal vesicle contains substances necessary for subsequent cleavage of the egg, these substances need not be present in the maturing oocyte during most, if not all, of the maturation period. And they obviously do not result from hormone action, as they are already present in the germinal vesicle prior to hormone exposure. Most likely these substances and others with specific influence on post-fertilization events (Briggs and Cassens, 1966) arise during oogenesis, are stored in the germinal vesicle and are released into and become a part of the cytoplasm at the time of germinal vesicle breakdown. "Chemical enucleation" as defined by Davidson (1968) occurs when the production (transcription) of primary gene products is inhibited by the presence of a chemical inhibitor, such as actinomycin D. Although such negative results have their obvious limitations, we have found that injection of oocytes with an amount of actinomycin D that will inhibit more than 80% of oocyte RNA synthesis (LaMarca, Ecker, and Smith, 1971) and will cause embryos to be arrested as blastulas, has no effect on hormone-stimulated oocyte maturation (Smith and Ecker, 1969). These observations are consistent with those obtained with oocytes enucleated by removal of the germinal vesicle; that is, the events of hormone induced maturation are solely under cytoplasmic control. In addition, the results with actinomycin D suggest that these events are not the result of transcription from cytoplasmic determinants during maturation. Thus, the most attractive conclusion is that any protein synthesis occurring in the oocyte during maturation is determined by stable templates, synthesized during oognesis and stored in the cytoplasm.

Protein synthesis in enucleated oocytes

Comparisons were previously made of amino acid incorporation kinetics during maturation between enucleated oocytes and their nucleated controls (Smith and Ecker, 1969). These comparisons showed no differences between the two groups of oocytes. However, because enucleation may substantially change the size of the amino acid pools and, as we saw in the preceding section, such pool size changes can have a marked effect on amino acid incorporation, specific rates of protein synthesis were determined for both enucleated and nucleated oocytes during hormone-induced maturation, using the technique outlined earlier in this report. The results of these determinations are shown in Table 1. These data show that, as reported earlier (Smith and Ecker, 1969), the rate of protein synthesis is low in oocytes that have not been exposed to hormone and rises significantly after hormone stimulation. Of particular importance is the fact that there does not appear to be a great difference in rate at any time during maturation between the enucleated oocytes and their nucleated controls. Certainly there is no indication that enucleation decreases the rate. Thus, although the pattern of rate increase after hormone exposure is somewhat different here than that inferred earlier from comparative values of β only (Smith and Ecker, 1969), the general conclusions remain the same. That is, protein synthesis in R. pipiens oocytes is accelerated following hormone exposure and the timing and amount of the increase is essentially the same whether or not the germinal vesicle was removed prior to hormone treatment.

It is apparent that at least some portion of the protein synthesis that occurs during maturation is immediately essential for the continuation of the process, because injection of puromycin (0.2 μg/oocyte) at any time in the period of maturation stops the process at that point (Smith and Ecker, 1969). In addition, if oocytes are incubated during the early stages of maturation under anaerobic conditions, their rate of protein synthesis is greatly diminished and they do not mature, even when later restored to aerobic conditions (Smith and Ecker, 1970). And, if the ionic content of the medium is adjusted in such a way that protein synthesis is inhibited, maturation will not occur (Ecker and Smith, 1971a). However, other evidence suggests that, in the later stages of maturation, the amount of this protein synthesis may be small. If, after first meiosis oocyte protein synthesis is decreased by anaerobic conditions, maturation can continue (Smith and Ecker, 1970).

Some of the proteins synthesized during maturation may not be essential to maturation, but are probably important in later development. These are the proteins that later concentrate in the cleavage nuclei (Ecker and Smith, 1971b). Although proteins synthesized early in maturation also concentrate in the germinal vesicle (Ecker and Smith, 1971b); synthesis of these proteins in the presence of the germinal vesicle does not appear to be necessary to later localization in the cleavage nuclei. We now find that proteins, synthesized in oocytes from which germinal vesicles were removed prior to hormone exposure, still accumulate rapidly in transplanted blastula nuclei after activation. Thus, in every aspect we have observed of protein synthesis and its control during maturation, the enucleated oocyte appears to be completely equivalent to its nucleated counterpart.

If the germinal vesicle plays no active role in the regulation of maturation after hormone exposure, what is its function in the oocyte ? It seems most likely that its active function ends at the end of oogenesis; and one line of evidence suggests that the hormone action that initiates the resumption of meiosis may also cause the termination of oogenesis. We have found that the action of progesterone on isolated oocytes in vitro causes a marked decrease in oocyte RNA synthesis (LaMarca, Ecker, and Smith, 1971). Particularly sensitive to hormone-induced inhibition is high-molecular-weight (possibly ribosomal) RNA, which is generally believed not to be synthesized between the end of oogenesis and the onset of gastrulation. Therefore we suggest that the germinal vesicle is essentially an "oogenesis nucleus," directing the formation of gene products which are localized in the oocyte cytoplasm and programmed to specific function during the time of "nuclear inactivity," that is, from the end of oogenesis to the beginning of gastrulation (see Ecker, Smith, and Subtelny, 1968; Ecker and Smith, 1971b).

We have little direct evidence to provide an insight into the mechanisms by which this non-nuclear regulation may occur. As it relates to the stimulation of protein synthesis following hormone exposure, we do know several things. First the increase in protein synthesis occurs only in oocytes which have received the hormonal stimulus and, second, the capacity of oocytes to synthesize protein (and to undergo normal maturation) is correlated with the ability of the oocyte to take up and retain potassium (Ecker and Smith, 1971a). Because of what is already known concerning the possible role of potassium in the regulation of protein synthesis (see Lubin, 1964; Molinaro and Hultin, 1965) it seems reasonable, then, that it is at this level that we should look for the controls regulating the hormone-mediated initiation of protein synthesis in the amphibian oocyte.

Concluding remarks

The observation made in this work that the amino acid pools active in protein synthesis may be orders of magnitude smaller than the amount of free amino acid extractable from the system, raises some questions concerning the accuracy of determinations of protein synthetic rate based solely on the size of extractable pools. Because of its relatively large size, the amphibian oocyte may provide a greater problem in this respect than other cell systems. However, the possibility of such discrepancies must be considered, whatever cell system is being used.

In the study presented here knowledge of the absolute values of protein synthetic rates is not as essential as a dependable comparison between nucleated and enucleated oocytes. Although we had earlier made such a comparison without consideration of absolute rates (Smith and Ecker, 1969), we now find that 10-20% of the total extractable free leucine is localized in the germinal vesicle before hormone exposure. Thus, it was essential in this study to show that, although enucleation removes a substantial fraction of the free leucine of the oocyte, it has little or no effect on either the size of the active leucine pool or the rate of protein synthesis in the oocyte.

Acknowledgments

This work was done with the technical assistance of Miss Gloria Kokaisl, whose contribution is gratefully acknowledged. Most of the computer programming was the work of Mrs. Jeanne Blomquist, whose assistance is also acknowledged with thanks. This work was done under the auspices of the U.S. Atomic Energy Commission.

References

Briggs, R.W. and Cassens, G. (1966) - Proc. Nat. Acad. Sci. U.S. 55 1103-1109.
Brown, D.D. and Dawid, I.B. (1968) - Science 160, 272-280.
Davidson, E.H. (1968) - "Gene Activity in Early Development." Academic Press, New York.
Dawid, I.B. (1965) - J. Mol. Biol. 12, 581-599.
Ecker, R.E. and Smith, L.D. (1966) - Biochim. Biophys. Acta 129, 186-192.
Ecker, R.E. and Smith, L.D. (1968) - Develop. Biol. 18, 232-249.
Ecker, R.E. and Smith, L.D. (1971a) - J. Cell. Physiol. 77, 61-70.
Ecker, R.E. and Smith, L.D. (1971b) - Develop. Biol. 24, 559-576.
Ecker, R.E., Smith, L.D., and Subtelny, S. (1968) - Science 160, 1115-1117.
LaMarca, M.J., Ecker, R.E., and Smith, L.D. (1971) - Develop. Biol. Submitted for publication.
Lubin, M. (1964) - Fed. Proc. 23, 994-1001.
Molinaro, M. and Hultin, T. (1965) - Exp. Cell. Res. 38, 398-411.
Smith, L.D. and Ecker, R.E. (1969) - Develop. Biol. 19, 281-309.
Smith, L.D. and Ecker, R.E. (1970) - Develop. Biol. 22, 622-

637.

Smith,L.D., Ecker,R.E., and Subtelny,S. (1966) - Proc. Nat. Acad. Sci. U.S. 56, 1724-1728.

Wallace,R.A. and Jared,D.W. (1969) - Develop. Biol. 19, 498-526.

Table 1.

Rate of protein synthesis in nucleated and enucleated oocytes at various times after hormone treatment.

Experiment[1]	Time after progesterone (hrs)	Type of oocytes	Stage of maturation at injection[2]	Rate of protein synthesis[3] (μg/oocyte/hr)
I	0	Nucleated	No maturation	< 0.005
		Enucleated	—	< 0.005
	23	Nucleated	1st meiotic metaphase	0.046
		Enucleated	—	0.037
II	0	Nucleated	No maturation	< 0.005
		Enucleated	—	< 0.005
	6	Nucleated	G.V. still intact	0.018
		Enucleated	—	0.018
	12	Nucleated	Early 1st metaphase	4
		Enucleated	—	0.020
	$16\frac{1}{2}$	Nucleated	1st meiotic metaphase	0.040
		Enucleated	—	0.048
	$20\frac{1}{2}$	Nucleated	1st polar body	0.014
		Enucleated	—	0.048
	24	Nucleated	Early 2nd meiosis (not yet metaphase)	0.020
		Enucleated	—	0.068

Table 1 (cont.)

Oocytes were removed from their ovarian follicles, treated with progesterone and, after the incubation times indicated in the table, were injected with 17.8 nl of tritiated 1-leucine at two different concentrations (0.0455 µmoles/ml and 0.198 µmoles/ml ; both had an activity of 1.0 mCi/ml). Double sets of kinetics were obtained as outlined in Figure 2 and the value of f (the rate of flow of leucine through the amino acid pool) was calculated from the resultant data using equations (1-3). The rate of protein synthesis was obtained from f assuming 10% of the protein mass is leucine. The calculated size of the active amino acid pool for all of these experiments varied between 2×10^{-6} and 6×10^{-6} µmoles/oocyte.

[1] Each group of experiments was with oocytes from a separate female.

[2] Stages of maturation vary somewhat among different animals and the process tends to be somewhat accelerated in frogs collected in the spring.

[3] In those experiments in which uptake was very slow (no hormone treatment), an upper limit of 0.005 µg/oocyte/hr was determined from the initial uptake kinetics, an estimate of α from other experiments in the series and an assumption of a pool size of 6×10^{-6} µmoles/oocyte, the maximum observed in any other experiment in the series (see Ecker and Smith, 1966; 1968).

[4] Inadequate data to compute this point.

ANUCLEATE SYSTEMS: BACTERIA AND ANIMAL CELLS

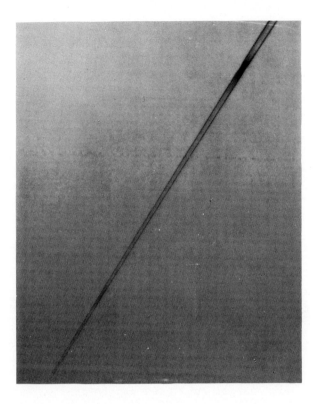

Fig. 1 Micropipette used for injecting oocytes. This pipette measures about 40 µm across at the tip and delivers 17.8 ± 0.3 nl between the two constrictions.

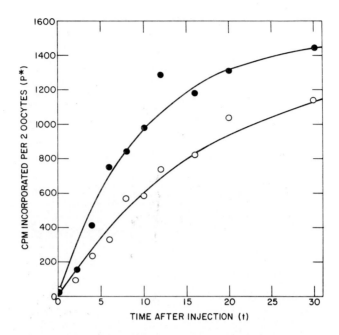

Fig. 2 Incorporation kinetics of microinjected 1-leucine-4,5-^{3}H (Amersham/Searle Corp., Des Plaines, Illinois) in maturing oocytes. Oocytes had been removed from their ovarian follicles and treated with progesterone (10 µg/ml, 10min.) 20 hrs. earlier. They were incubated in amphibian Ringer's solution at 18-19°C during that time and were at about first meiotic metaphase at the time of injection. Each oocyte was injected individually with 12.5 nl of the labelled amino acid; closed circles 0.100 mCi/ml and 0.100 µmoles/ml - open circles, 0.091 mCi/ml and 0.230 µmoles/ml. At the indicated times after injection, two oocyte samples were taken and prepared as outlined in the text. The curves shown in the figure are the best fit by computer of equation (1) to each set of experimental data.

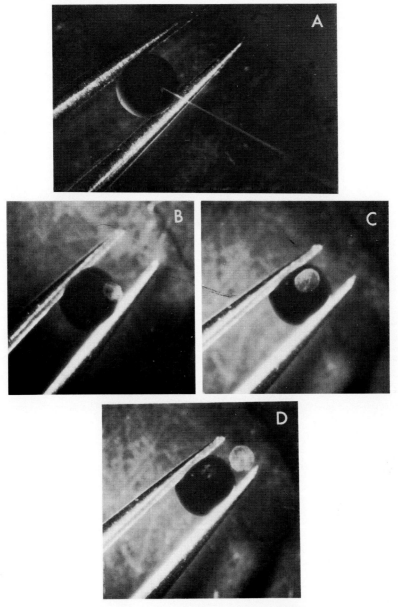

Fig. 3 Enucleation of a full-grown *Rana pipiens* oocyte. A. Holding the oocyte with watchmaker's forceps, a short incision is made at the animal pole with a glass needle. B-D. Applying gentle pressure on the oocyte with the forceps, the nucleus (germinal vesicle) begins to emerge and, with continued pressure, emerges completely and falls to the bottom of the dish.

DNA, RNA AND PROTEIN SYNTHESIS IN ANUCLEATE FRAGMENTS OF SEA URCHIN EGGS[1]

Sydney P. Craig[2]

Developmental Biology Branch
National Institute of Child Health and Human Development
National Institutes of Health
Bethesda, Maryland 20014, U.S.A.

and

Division of Biology
California Institute of Technology
Pasadena, California, U.S.A.

Abstract

DNA, RNA and protein syntheses were examined in artificially activated anucleate fragments of eggs of the sea urchin, Strongylocentrotus purpuratus. Most of the RNA synthesized by anucleate fragments within 1 h after activation has less secondary structure than ribosomal RNA when examined by benzoylated-DEAE-cellulose chromatography. This new RNA was associated entirely with the 15,000 x g pellet of the homogenate of the fragments.

---

1. Requests for reprints should be sent to the National Institute of Child Health and Human Development, National Institutes of Health, Bethesda, Maryland 20014, U.S.A.
2. A portion of this work was submitted to the California Institute of Technology in partial fulfillment of the requirements for the degree of Doctor of Philosophy.

Artificially activated anucleate fragments synthesized proteins which associated with mitochondria which were purified by sucrose buoyant density gradient centrifugation. This synthesis was unaffected after 6 h stringent inhibition of the RNA synthesis with ethidium bromide. However, ethidium bromide inhibited the synthesis of protein in the 100,000 x g supernatant fraction by approximately 39 %. The cytoplasmic proteins which migrated with a molecular weight of 10,000 to 20,000 in sodium dodecyl sulfate-polyacrylamide gels were especially affected by ethidium bromide, being inhibited by more than 53 %.

Tritiated thymidine was incorporated into DNA in artificially activated anucleate fragments. The results from several methods of investigation suggest that mitochondrial DNA was not the template for the DNA synthesis. The results from these investigations include : a) the molecular weight of 7.6×10^5, calculated from the width of the band of DNA at equilibrium in CsCl ; b) the failure of ^{3}H-thymidine to be incorporated into closed circular mitochondrial DNA as prepared by CsCl-ethidium bromide buoyant density centrifugation ; c) the failure of ^{3}H-thymidine to become incorporated into the 27 S nicked mitochondrial DNA in isolated mitochondria ; d) the failure to discover replicative forms among the non-closed (nicked) circular mitochondrial DNA through preliminary electron microscopic investigations.

The use of uridine and thymidine as precursors to nucleic acid synthesis in the above studies led to the demonstration that uridine is converted to cytidine in artificially activated anucleate fragments and that the polysaccharide fraction becomes labeled during extended pulses with ^{3}H-thymidine.

Introduction

The presence of DNA in mitochondria (for review see Piko et al., 1967) and the possible existence of other forms of cytoplasmic DNA leads one to inquire as to their contribution to biosyntheses within cells, and in the case of the sea urchin, to their contributions to the developing embryo. The anucleate fragment of the sea urchin egg provides a convenient system for examining the synthesis of nucleic acids and proteins in the cytoplasm without the interference of nuclear activity. Nucleic acid and protein synthesis in anucleate fragments is not necessarily related to cytoplasmic DNA, however. Indeed, the vast majority of the proteins synthesized in anucleate fragments are translated from templates which were present in the unfertilized egg (for references see Tyler, 1966 ; Craig & Piatigorsky, 1971). On the other hand, at least some of the RNA synthesized by anucleate fragments (Craig, 1970) and by blastula and gastrula stage embryos (Hartman & Comb , 1969) of the sea urchin has been shown to be homologous to the mitochondrial DNA template. RNA synthesis in anucleate fragments has also been studied by Baltus et al. (1965), Selvig et al. (1970) and Chamberlain (1970). Recent attempts to identify proteins synthesized upon mitochondrial RNA templates have been unsuccessful (Craig & Piatigorsky, 1971). Thus, at present, the function of mitochondrial RNA's is not known.

In the present study, RNA and protein synthesis by anucleate fragments were examined further. Particular emphasis was placed on locating the newly synthesized RNA within the fragments. Protein synthesis was examined in relation to the synthesis and localization of RNA in the artificially activated anucleate fragments.

DNA synthesis, by anucleate fragments, was also examined. The results of these investigations are discussed with particular emphasis on the possibility that cytoplasmic DNA, other than mitochondrial DNA, may be replicated during early development.

Material and Methods

Handling and fragmentation of eggs

Sea urchins of the species <u>Strongylocentrotus purpuratus</u> were collected personally and commercially (Pacific Bio-Marine Supply Company) from the coast of southern California. The sea urchins were maintained in the laboratory and were periodically induced to spawn by an injection of 1 ml of 0.55 M KCl (see Tyler & Tyler, 1966). The shed eggs were strained through cheesecloth and washed with artificial sea water. The gelatinous coat was removed by washing with artificial sea water at pH 4.5.

Unfertilized eggs were fragmented into nucleate and anucleate fragments by centrifugation in buoyant, isotonic, sucrose-sea water gradients following the basic procedures outlined elsewhere (Harvey, 1956 ; Tyler, 1966). The primary modification was to change the centrifugation from 5min at 2,500 rpm followed by 15 min at 10,000 rpm (in a Spinco S.W. 25.1 rotor) to 5 min at 2,000 rpm followed by 30 min at 12,000 rpm. The anucleate fragments were collected after puncturing the bottom of the centrifuge tube and were washed with ice cold sea water. Representative samples of fragments were fixed and dehydrated and cleared by xylene treatment ; no nuclei were visible. The fragments were artificially activated by treatment with 0.005 M butyric acid at 18°C for 1.0 - 1.75 min (Tyler, 1966). For studies with embryos, eggs were fertilized and cultured in artificial sea water, as described by Tyler & Tyler (1966). Incubations were terminated by washing the cells 3 times by centrifugation (500 x g) in ice cold artificial sea water.

Homogenization and fractionation

Anucleate fragments were homogenized with 5 to 10 strokes in a Dounce homogenizer in 4 volumes of 0.45 M KCl, 0.02 M Mg acetate, 0.05 M Tris at pH 7.6 containing 1 mg/ml of purified bentonite. The homogenate was centrifuged at 15,000 g for 20 min at 4°C. The supernatant fractions and the pellets were analyzed, as given with the individual experiments.

For isolation of mitochondria, fragments were homogenized in 5 volumes of 0.9 M sucrose, 0.003 M ethylenediamine tetraacetic acid (EDTA) and 0.003 M Tris at pH 7.6 and were layered onto a 0.93 M to 1.88 M sucrose density gradient in the same buffer. The gradients were centrifuged at 40,000 rpm for 40 min in an S.W. 41 rotor at 2°C, as modified from Shaver (1956) and Piko et al. (1967). After collection, the fractions comprising the initial sample layer, minus the yolk, were combined and labeled the 100,000 x g supernatant fraction. The fractions containing the isopycnically banded mitochondria were diluted with 4 volumes of 0.5 M KCl, 0.005 M EDTA and 0.05 M Tris at pH 7.6 and centrifuged at 15,000 x g for 20 min in order to pellet the mitochondria.

Nucleic acid extraction

DNA or RNA was extracted after suspending the washed, labeled fragments in 4 volumes of ice cold buffer (0.005 M EDTA, 0.275 M KCl, and 0.05 M Tris at pH 7.5 for DNA or 0.01 M NaCl, 0.01 M NaAc at pH 5 for RNA). Bentonite was added to 1 mg/ml and sodium deoxycholate (Na DOC) was added to a final concentration of 0.5 %. After the suspension cleared, SDS was added to 2 %, forming a precipitate which dissolved after warming to room temperature. The resulting solution was then extracted with an equal volume of phenol containing 0.1 % 8-hydroxyquinoline. The extractions were conducted at room temperature, with constant mixing, for 30 min. After each extraction the mixture was centrifuged at 5,000 rpm at 4°C in a Sorvall Swinging Bucket Rotor and the aqueous phase was removed. After the second phenol extraction, the nucleic acid was precipitated from the aqueous phase overnight at -20°C after the addition of 2 volumes of ethanol. The precipitate was washed 3 x in 100 % ethanol and 1 x in 100 % ether and dried. Between washes the precipitate was repelleted by centrifuging at 5,000 x g for 10 min at 4°C. The dried pellets were then ready to be dissolved in the appropriate buffer for subsequent analysis.

Benzoylated - diethylaminoethyl cellulose (B-D-cellulose) chromatography.

The B-D-cellulose chromatography followed the procedures of Sedat et al. (1969), with the exception that an increasing gradient of deionized urea was included in the salt gradients in order to force the nucleic acids to elute at lower salt concentrations than would otherwise occur.

TCA precipitation of nucleic acids for radioactive analysis

Aliquots of the fractions to be analyzed were dried on filter papers which were then washed 5 x with ice cold 5 % TCA, twice with 95 % ethanol, twice with absolute alcohol, and dried from anhydrous ethyl ether, as given by Mans & Novelli (1961) and modified by Tyler (1966). The papers were analyzed for radioactivity, after immersion in a

Toluene, POP, POPOP scintillation fluid, at approximately 6 % counting efficiency, for ^{3}H, in a Beckman Scintillation Spectrometer at ambient temperature.

Preparation of proteins for electrophoresis in polyacrylamide gels.

Isolated mitochondria and 100,000 x g supernatant fractions from cellular homogenates were heated to 100°C for 15 min in 8 % TCA and then refrigerated at 4°C for 16 h. The TCA extracts were centrifuged at 1,500 x g for 10 min at 4°C and the resulting pellets were washed once in 5 % TCA, twice in 95 % ethanol, once in 100 % ethanol and were dried from anhydrous ethyl ether. The extracted proteins were solubilized in 0.2 ml of 0.1 M Na phosphate 1 % SDS, and 1 % 2-mercaptoethanol (2-Me), pH 7.1, at 37°C for 3 h and were subsequently dialyzed against 3 changes of 0.01 M Na phosphate, 0.1 % SDS and 0.1 % 2-Me, pH 7.1, at room temperature, as given elsewhere (Shapiro et al., 1967). The solubilized proteins were electrophoresed in a composite gel of 4 % acrylamide and 0.5 % agarose, buffered with 0.1 M Na phosphate and 1 % SDS at pH 7.1 in a vertical slab apparatus (E - C Apparatus Corporation). Electrophoresis was at 200 mA (approximately 100 V) for 3 1/3 h at 20°C. Trypsin (Worthington Biochemical Corporation), bovine serum albumin and ovalbumin (Sigma Chemical Company) were examined in parallel gels as molecular weight markers. The proteins were stained with 0.25 % Coomassie Brilliant Blue (Shapiro et al., 1967). 1 mm slices were dissolved overnight in 30 % hydrogen peroxide and aired for several hours to reduce the background levels in subsequent scintillation counting. A toluene based scintillation coctail with approximately 3.3 % liquifluor and 8.5 % biosolv - 3 (from Beckman Instruments) was added to each sample and the radioactivity was determined at better than 90 % counting efficiency.

CsCl buoyant density centrifugation

In experiments where DNA was analyzed by CsCl buoyant density centrifugation, the fragments were first dissolved in 4 volumes of a buffer containing 4 % SDS, 9 % ethanol and 0.08 M EDTA at pH 8. The resulting cell lysate was adjusted to 1.7 gms/cc with CsCl and centrifuged at 40,000 rpm for 48 h at 20°C in a S.W. 65 rotor (Piko et al.,1967). Two tenths ml fractions were collected. In some experiments the material banding at 1.7 gms/cc was diluted to 1.55gms/cc, made to 300 µgs/ml in ethidium bromide and recentrifuged in the S.W. 65 rotor at 40,000 rpm for 48 h at 20°C (Piko et al., 1968). Fluorescence of the rebanded DNA under UV light permitted visual localization of the bands of DNA

The redistribution of label as examined through base analysis.

The nucleic acid was extracted from activated anucleated fragments, labeled for 6 h with ^{3}H-5-uridine, as described above. The dried nucleic acid was dissolved in 1.0 ml of 0.275 M KCl, 0.005 M EDTA and 0.05 M Tris, pH 7.5, and treated with 10 µg of RNAase (preheated to 100°C for 1 min) for 1 h at 25°C. The remaining nucleic acid was precipitated with 2 volumes of ethanol at -20°C for 4 h and was subsequently washed twice in ice cold 5 % TCA before dehydrating and drying. The remaining nucleic acid was then hydrolyzed and column chromatographed through a Dowex 50 - X8 resin according to the procedures of Brown & Attardi (1965) in order to separate adenine, guanine, and cytosine from uridine and thymidine.

Electron microscopy

The DNA from fertilized and unfertilized eggs was purified by CsCl ethidium bromide buoyant density centrifugation, as given above. This DNA was prepared directly for electron microscopy, according to Methods described previously (Piko et al., 1968 ; Kleinschmidt & Zahn, 1959).The electron microscope grids were shadowed with 80 % platinum 20 % palladium for 1 min while rotating constantly (approximately 5 rpm). The grids were examined and photographed in an RCA EMU 3-D electron microscope.

Source of Chemicals

Ethidium bromide (Calbiochem) Uridine-5-^{3}H, $\geqslant$ 20 Ci/mmole ; L-valine-^{14}C, 143 mCi/mmole ; thymidine-met-^{3}H, $\geqslant$ 15 Ci/mmole ; and B-D-cellulose, 50 - 500 mesh (Schwarz BioResearch Corporation). Cesium chloride (Harshaw Chemical Company). Dowex 50 - X8 (Bio-Rad Laboratories). 80 % platinum, 20 % palladium wire (Ernest F.Fullam,Inc.). Sucrose (ribonuclease-free, density gradient grade) and Coomassie Brilliant Blue R-250 (Mann Research Laboratories). Seakem Agarose (Marine Colloid, Inc.). Acrylamide ; N,N-methylenebisacrylamide ; and N,N,N',N'-tetramethylethylenediamine (Eastman Organic Chemicals). Ribonuclease A ; SDS and 2-Me (Sigma Chemical Company).

Results

RNA synthesis by artificially activated anucleate fragments of sea urchin eggs.

A previous study showed that artificially activated anucleate fragments synthesized RNA which had little secondary structure relative to the extra mitochondrial ribosomal RNA and which was at least partially homologous to mitochondrial DNA (Craig, 1970). In the present investigations, in order to localize the newly synthesized RNA, labeled anucleate fragments were homogenized and centrifuged at 15,000 x g to sediment yolk, mitochondria and membranes. The newly synthesized RNA (from 1 h labeling period), as analyzed by B-D-cellulose chromatography, was found associated almost exclusively with the sediment of this low speed centrifugation (Fig.1).

In other experiments, newly synthesized RNA could be shown to be associated with mitochondria isolated from fragment homogenates.

As uridine was utilized in these experiments as a precursor to RNA synthesis, the possibilities for its conversion to other nucleotides were investigated. The nucleic acid, extracted from anucleate fragments labeled for 6 h with uridine-5-^{3}H, was treated with ribonuclease A before washing in cold acid and analyzing for base composition (Fig.2). Under these conditions, DNA and double stranded RNA such as transfer RNA would be relatively insensitive to the ribonuclease treatment. The results from the chromatographic separation of the bases, illustrated in Fig.2, show that tritium label was shifted from uridine to cytidine in anucleate fragments. The labeled hydrogen probably remained in the 5 position during this conversion (Mahler & Cordes, 1966). Labeling of transfer RNA may, therefore, result from the turnover of the CCA terminus, rather than from its biosynthesis (Malkin et al., 1964; Gross et al., 1965). The potential for labeling of DNA in anucleate fragments, therefore, must also be considered in experiments using uridine as precursor.

Effect of ethidium bromide on the synthesis of protein within artificially activated anucleate fragments.

The proteins, associated with isolated mitochondria, which were synthesized in the absence of mitochondrial RNA synthesis were investigated in hopes of identifying specific proteins which may have been translated from the mitochondrial RNA. The pattern of proteins labeled for 30 min, 6 h after artificial activation of anucleate fragments, was investigated by SDS-polyacrylamide gel electrophoresis and the results were compared with those obtained after treatment of the fragments for 6 h with ethidium bromide, previously shown to be effective in inhibiting RNA synthesis by anucleate fragments (Craig & Piatigorsky, 1971). In these experiments, only those proteins associated with the isolated mitochondrial and the 100,000 x g supernatant fractions of fragment homogenates were analyzed (Fig.3). The results show that the synthesis of those proteins associated with subsequently isolated mitochondria is totally unaffected by inhibition of RNA synthesis during the previous six hours (Fig.3 A). On the other hand, ethidium bromide inhibited incorporation of label into protein by an average of 38 % in the 100,000 x g supernatant fraction. The radioactivity was decreased by more than 53 % in the proteins which migrated with a molecular weight corresponding to 10,000 to 20,000 (fraction numbers 70 to 78, Fig. 3 B).

DNA synthesis in artificially activated anucleate fragments of sea urchin eggs

Artificially activated anucleate fragments incorporated ^{3}H-thymidine into DNA while non-activated fragments did not (Fig.4). Although no radioactivity was incorporated into acid precipitate DNA, by the washed non-activated anucleate fragments, more than 12,000 acid soluble counts per min (as counted on filter papers) were retained by them. On the other hand, of the 60,000 counts per min accumulated by the activated anucleate fragments, more than 4,000 counts per min were incorporated into acid precipitable material with an average buoyant density of 1.7 gm/cc which is characteristic of DNA.

The profile of the band of radioactivity from the activated fragments can be correlated to the homogeneity of the DNA molecules and to the apparent molecular weight of the labeled DNA (Meselson et al., 1957). The downward concavity of the leading and trailing edges of the band of radioactivity is evidence for homogeneity in buoyant densities of the labeled molecules. Assuming that most of the synthesized DNA is a homogeneous species and utilizing the equation for molecular weight ($M = RT/\bar{v} \, (d\rho/dr)_r \, w^2 r_o \, \sigma^2$) given by Meselson et al. (1957) with the density at band center (1.7 gm/cc) as an approximation of the partial specific volume and with the calculated statistical deviation for the band width ($\sigma = 0.677$ cm), the apparent molecular weight of the DNA in the radioactive band is 7.6×10^5.

The DNA was rebanded in CsCl after intercalation with ethidium bromide in order to test for ^3H label in the closed circular mitochondrial DNA (Fig.5). After equilibrium, the DNA bands fluoresced orange in ultra violet light and the preparation fluoresced pale blue. The fractions labeled a, b and c (Fig.5) designate the positions for the visible bands of polysaccharide, closed circular mitochondrial DNA and the non-closed DNA, respectively. After 2 h (Fig.5 A) and 8 h (Fig.5 B), incubation with ^3H-thymidine, incorporation into closed circular mitochondrial DNA was not found in significant quantities in either fertilized nucleate fragments or in activated anucleate fragments. The proportion of closed circular DNA in anucleate fragments was not visibly decreased after artificial activation as might have been expected if a large proportion of the mitochondrial DNA were stimulated to replicate.

Curiously, considerable radioactivity became associated with the polysaccharide fraction after 8 h incubation with ^3H-thymidine in anucleate fragments, but not in embryos.

When ^3H-thymidine labeled anucleate fragments were phenol extracted and the nucleic acid was analyzed by sucrose sedimentation (Fig.6), radioactivity was found to sediment around 38 and 27 S. Once again, none of the labeled DNA taken either before or after sucrose sedimentation was found to band with the closed circular mitochondrial DNA after intercalation and rebanding in CsCl.

When mitochondria were isolated from artificially activated anucleate fragments, labeled for 1 h with ^{3}H-thymidine, and were subsequently lysed and analyzed by sucrose sedimentation, no radioactivity was observed to be incorporated into macromolecules sedimenting faster than about 10 S (Fig.7). Similar results were obtained in experiments where the total sediment of a 15,000 x g centrifugation of fragment homogenates was lysed and analyzed. A peak of ultra violet light absorption at approximately 27S (relative to 50 and 30 S ribosomal subunits in a parallel tube) was shown in other experiments to contain material with a buoyant density (1.70 gm/cc) and alkaline stability similar to that for DNA. This, plus the fact that the nicked monomer of mitochondrial DNA sediments at 27 S in sucrose gradients (Piko et al., 1968), suggests that the 27 S material observed in the mitochondrial lysates is the nicked monomer for mitochondrial DNA.

As replication of the mitochondrial DNA would be expected to require at least a single strand nick, the total non-closed circular band of DNA (including the nicked circular DNA) from fertilized embryos, was examined electron microscopically in search of replicative forms. Although the incidence of circles among the non-closed circular DNA from fertilized eggs was very low, a few circles were present and were found to be approximately 4.5 microns in length. None of the circles observed possessed tails or branches as might have been expected among replicating circular DNA's (Kiger & Sinsheimer, 1971).

Discussion

DNA, RNA and protein are synthesized in artificially activated anucleate fragments of sea urchin eggs. At least some of the RNA synthesized is transcribed from mitochondrial DNA (Craig, 1970) and, after a 1 h labeling period, it is found associated with the low speed sediment of cellular homogenates. Since this sediment contains the large particulate matter of the cell, such as the mitochondria, membranes and yolk, it may be concluded that the newly synthesized RNA remains membrane or organelle bound during the first hour after its synthesis.

The present results confirm and extend the earlier observations (Craig & Piatigorsky, 1971) that inhibition of mitochondrial RNA synthesis with ethidium bromide does not affect the labeling of proteins associated with mitochondria. Possible explanations for this include : a) The vast majority of protein synthesis in artificially activated a-nucleate fragments is upon stable templates which were present in the unfertilized egg, and conversely, mitochondrial RNA synthesis is at best a minor contributor to protein synthesis after activation. b) Mitochondrial RNA is not translated into protein. c) Mitochondrial RNA is translated at a place and/or time remote from its synthesis. Inhibition of RNA synthesis with ethidium bromide, however, did reduce the labeling of proteins present in the 100,000 x g supernatant fraction by more than 38 %. One of the more interesting possibilities is that this reflects the participation of mitochondrial RNA in protein synthesis outside of the mitochondria. On the other hand, ethidium bromide may also affect the turnover of cytoplasmic proteins more than that of mitochondrial proteins, or may simply interfere with one or more of the steps of protein synthesis in the cytoplasm more effectively than in the mitochondria.

The failure to demonstrate the association of new DNA with mitochondrial DNA in activated anucleate fragments is unexpected and problematical. About 80 % of the DNA from unfertilized eggs has been reported to be cytoplasmic (Piko et al., 1967). The exact proportion of that which is mitochondrial is not known, although it was suggested that yolk may also possess DNA (Piko et al., 1967). However,the DNA in yolk could be contaminants from mitochondria.

There is evidence against the replication of mitochondrial DNA during early cleavage of mouse and sea urchin embryos (Piko, 1970). This was based on the fact that ^{3}H-thymidine was not incorporated into closed circular mitochondrial DNA. The results reported here are also consistant with mitochondrial DNA not being synthesized during early cleavage. The present evidence for this is : a) The DNA synthesized by anucleate fragments was not found associated with the 27 S material, containing mitochondrial DNA, in lysates of isolated mitochondria. b) The proportion of closed circular DNA in anucleate fragments was not visibly depleted after artificial activation. c) The calculated average molecular weight for the synthesized DNA was 7.6×10^5 versus 1.4×10^7 for mitochondrial DNA of sea urchins (Piko et al., 1967). d) Replicative forms were not observed in the electron microscope among the non-closed (nicked) circular mitochondrial DNA. e) ^{3}H-thymidine was not incorporated into closed circular mitochondrial DNA.

On the other hand, the synthesis of mitochondrial DNA in anucleate fragments or during early cleavage cannot be excluded on the present evidence. For example, the calculation of molecular weight appearing in c) above was based on a formula which depended upon measurement of the band width at equilibrium in CsCl. In the preparative ultracentrifuge, the band width may be affected by rotor deceleration and fraction collecting procedures. Furthermore, ^{3}H-thymidine labeled 37 and 27 S material in phenol extracts of anucleate fragments could be composed of doubly nicked catenate dimers and singly nicked monomers of mitochondrial DNA, as described elsewhere (Brown & Vinograd, 1971). And, in addition, electron microscopic autoradiographic data has also been reported which suggests that the mitochondrial DNA of sea urchins replicates during early cleavage (Anderson, 1969).

However, the experiments suggest that cytoplasmic DNA, other than mitochondrial DNA is replicated after artificial activation of anucleate fragments. Centrioles have been observed to replicate in artificially activated anucleate fragments of sea urchin eggs (Harvey, 1956). Other organisms, such as fungi, have been reported to possess DNA in the centrioles (Weijer, 1965 ; McDonald & Weijer, 1966). Thus, centriolar DNA may be supplying the template for the synthesis of DNA. Another possible explanation for the DNA synthesis in anucleate fragments could involve the leakage and subsequent replication of nuclear DNA into the cytoplasm during or before fragmentation.

The incorporation of label into polysaccharide during ^{14}C-thymidine labeling of mouse embryos has been reported (Piko, 1970). The results of the experiments reported here suggest that the incorporation of thymidine label into the polysaccharide fraction is not a normal consequence of early sea urchin development. Rather, it may result from a reduced demand on the thymidine pool, available for DNA synthesis in artificially activated anucleate fragments. Further experiments are necessary to resolve this point.

Acknowledgements

I am very grateful to the late Dr. Albert Tyler, with whom this work was initiated. I would also like to thank Drs. Joram P. Piatigorsky, Jerome Vinograd, Robert L. Sinsheimer and Joel A. Huberman for advice and assistance, Frieda Sobell for typing the manuscript and Paul K. Cartier for technical contributions.

References

Anderson, W.A., J.Ultrastructure Res., 26, 95 (1969).

Baltus, E., Quertier, J., Ficq, A. & Brachet, J., Biochim. Biophys. Acta, 95, 408 (1965).

Brown, G.M. & Attardi, G., Biochem. Biophys. Res. Commun., 20, 298 (1965).

Brown, I.H. & Vinograd, J., Biopolymers, in press (1971).

Chamberlain, J.P., Biochim. Biophys. Acta, 213, 183 (1970).

Craig, S.P., J.Mol.Biol., 47, 615 (1970).

Craig, S.P. & Piatigorsky, J., Develop. Biol., 24, 214 (1971).

Dahoff, M.O. & Eck, R.V., "Atlas of Protein Sequence and Structure" National Biomedical Research Foundation, Silver Spring, Maryland, (1967-1968).

Gross, P.R., Kraemer, K. & Malkin, L.I., Biochem. Biophys. Res. Commun., 18, 569 (1965).

Hartman, J.F. & Comb, D., J.Mol.Biol., 41, 155 (1969).

Harvey, E.B., "The American Arbacia and Other Sea Urchins" Princeton Univ. Press, Princeton, New Jersey (1956).

Ifft,J.B., Voet,D.H. & Vinograd,J., J.Phys.Chem., 65, 1138 (1961).

Kiger,J.A.,Jr. & Sinsheimer,R.L., Proc.Natl.Acad.Sci.U.S.A. 68, 112 (1971).

Kleinschmidt,von A. & Zahn,R.K., Z.Naturforsch., 14b, 770 (1959).

Mahler,H.R. & Cordes,E.H., In "Biological Chemistry" 728 Harper and Row, New York (1966).

Malkin,L.I., Gross,P.R. & Romanoff,P., Develop.Biol., 10, 378 (1964).

Mans,R.J. & Novelli,D.G., Arch.Biochem.Biophys., 94, 48 (1961).

McDonald,B.R. & Weijer,J., Can.J.Genet.Cytol., 8, 42(1966)

Meselson,M., Stahl,F.W. & Vinograd,J., Proc.Natl.Acad.Sci. U.S.A., 43, 581 (1957).

Piko,L., Develop.Biol., 21, 257 (1970).

Piko,L., Tyler,A. & Vinograd,J., Biol.Bull., 132, 68(1967)

Piko,L., Blair,D.G., Tyler,A. & Vinograd,J., Proc.Natl. Acad.Sci.U.S.A., 59, 838 (1968).

Sedat,J., Lyon,A. & Sinsheimer,R.L., J.Mol.Biol., 144, 415 (1969).

Selvig,S.E., Gross,P.R. & Hunter,A.L., Develop.Biol., 22, 343 (1970).

Shapiro,A.L., Vinuels,E. & Maizel,J.U.,Jr., Biochem.Biophys.Res.Commun., 28, 815 (1967).

Shaver,J.R., Exp.Cell Res., 11, 548 (1956).

Tyler,A., Biol.Bull., 130, 450 (1966).

Tyler,A. & Tyler,B.S., In "Physiology of Echinodermata" (Ed. R.A.Boolootian), Wiley, New York, p.639 (1966).

Weijer,D.L., Can.J.Genet.Cytol., 6, 383 (1964).

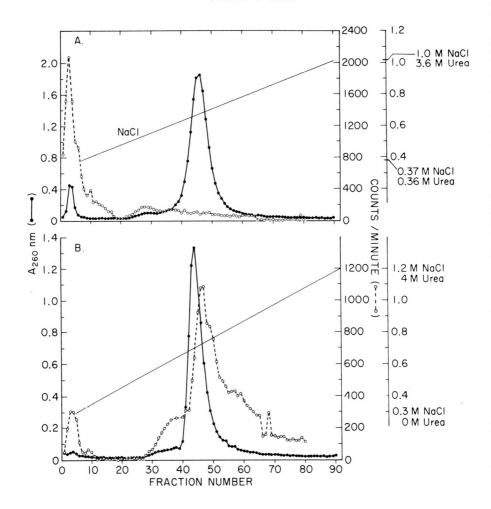

Fig. 1 - Benzoylated DEAE-cellulose chromatography of the nucleic acid extracted from the 15,000 x g Pellet (graph B) and the corresponding supernatant fraction (graph A) of an homogenate of anucleate fragments.

Anucleate fragments were incubated during the first hour after artificial activation with ^{3}H-5-uridine at 15 µCi/ml in a total volume of 30 ml and were then washed, homogenized, centrifuged and extracted with phenol, as specified in Materials and Methods. The alcohol precipitate was dissolved in 0.3 M NaCl, 0.001 M EDTA and 0.01 M Tris, at pH 5.6, adsorbed to B-D-cellulose equilibrated in the same buffer and eluted with an increasing gradient of NaCl and urea. The concentration of buffer (0.001 M EDTA and 0.01 M Tris, at pH 5.6) remained constant. Two ml fractions were collected. The light absorption at 260 nm was examined in a Beckman DU Spectrophotometer and 0.4 ml aliquots were TCA precipitated for radioactive counting, as specified in Methods. DNA and transfer RNA elute on the leading edge and messenger RNA on the trailing edge of the prominent absorption peak of ribosomal RNA present in fractions 40 - 50. The radioactivity which did not adsorb to the column (fractions 1 - 8) is assumed not to be nucleic acid (Sedat et al., 1969).

A_{260} nm •——•
Radioactivity o-----o
NaCl and urea gradient ———

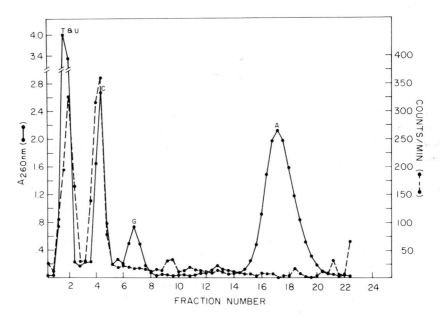

Fig. 2 - ^{3}H-5-uridine conversion to ^{3}H-cytidine; anucleate fragments of sea urchin eggs.

Artificially activated anucleate fragments were incubated for 6 hours at 19°C in 10 ml of artificial sea water with 15 µCi of ^{3}H-5-uridine per ml, after which they were washed and phenol extracted, as specified in Methods. The dried nucleic acid was dissolved in 1 ml of 0.275 M KCl, 0.005 M EDTA and 0.05 M Tris at pH 7.5 and treated with 10 µg of preheated RNAase for 1 hour at 25°C. The remaining nucleic acid was precipitated in 2 volumes of ethanol at -20°C for 4 hours and subsequently washed 2 times in ice cold 5 % TCA before rehydrating and drying. The remaining nucleic acid was hydrolyzed and column chromatographed through a Dowex 50-X8 column following the procedures of Brown & Attardi (1965). The bases elute in the order U & T, C, G and A (from left to right).

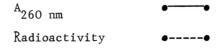

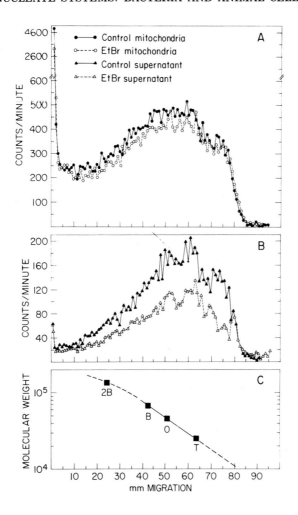

Explanation for Fig. 3 on following page.

Fig. 3 - SDS-polyacrylamide gel electrophoresis of protein synthesized by anucleate fragments in the presence and absence of ethidium bromide.

A suspension of activated anucleate fragments was divided into 2 parts, one was treated with 10 µg/ml of ethidium bromide and both were incubated at 17°C for 6 hours in a total volumn of 20 ml. The fragments were then labeled for 30 minutes with 0.125 µCi/ml of valine-^{14}C. Subsequently, the mitochondrial and 100,000 x g supernatant fractions were prepared and the proteins were extracted and analyzed, as specified in Materials and Methods.
Graph (A) mitochondrial proteins ;
Graph (B) supernatant proteins;
Graph (C) a calibration of millimeters of migration versus molecular weight based upon standard marker proteins. Molecular weights for bovine serum albumin (B), and ovalbumin (O) are from Shapiro et al., (1967) and for trypsin (T) from Dayhoff and Eck (1967-1968).
The dimer for bovine serum albumin is designated as 2 B.
The radioactivity in the ethidium bromide treated fractions was equated to the actual radioactivity in the control fractions utilizing, as a correction factor, the relative quantity of protein as estimated from the total binding of Coomassie Blue (measured by light absorption at 570 nm).

ANUCLEATE SYSTEMS: BACTERIA AND ANIMAL CELLS

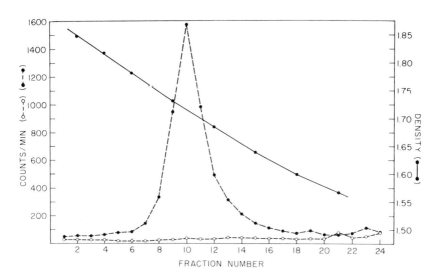

Fig. 4 - DNA synthesis in activated and non-activated anucleate fragments of sea urchin eggs.

Equal packed volumes (0.3 ml) of activated and non-activated anucleate fragments were incubated for 2 hours in 15ml of artificial sea water at 18°C with 13 µCi/ml ^{3}H-thymidine/ml. In this experiment the washed labeled fragments were suspended in 3 volumes of 0.275 M KCl, 0.01 M MgAc and 0.01 M Tris (pH 7.5) and were dissolved with 0.5 % Na-DOC. The resulting solution was made to 1.7 gms/cc with CsCl and centrifuged for 60 hours at 40,000 rpm at 20°C in a S.W. 65 rotor. 0.2 ml fractions were collected. Solution density was calculated from the refractive index (Ifft et al., 1961). Acid precipitable radioactivity ; non-activated fragments (o-----o) ; activated fragments (●-----●).

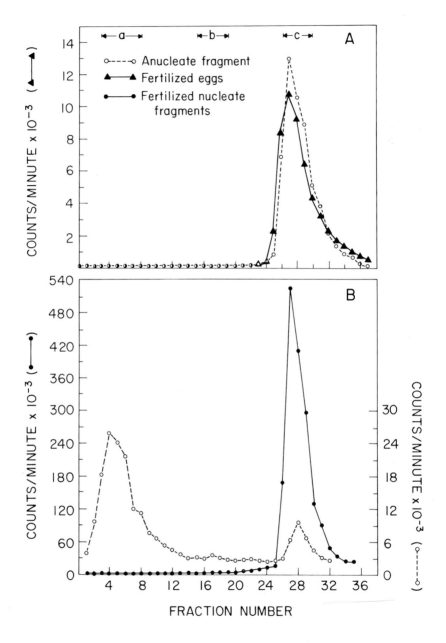

Explanation for Fig. 5 on following page.

Fig. 5 - CsCl buoyant density centrifugation of ^{3}H-thymidine labeled DNA intercalated with ethidium bromide.

Fertilized eggs (▲——▲) and artificially activated anucleate fragments (o-----o) were incubated for 2 hours at 18°C in 20 ml of artificial sea water containing 100 units/ml of penicillin and 20 µCi/ml of ^{3}H-thymidine (Graph A). In Graph B, fertilized nucleate fragments (●——●) and artificially activated anucleate fragments (o-----o) were incubated for 8 hours at 18°C in 40 ml of artificial sea water containing 100 units/ml of penicillin, 50 µgs/ml of streptomycin sulfate and 12.5 µCi/ml of ^{3}H-thymidine.
At the end of the incubation periods, the fertilized eggs (graph A) were at the two cell stage and the fertilized nucleate fragments (graph B) were at 4, 8, 16 and 32 cell stages. The DNA was prepared and banded in CsCl with ethidium bromide, according to the procedures outlined in Materials and Methods. The counts/minute represent the acid precipitable radioactivity. The fraction numbers designated by a, b and c are the fractions containing polysaccharide, closed circular mitochondrial DNA and non-closed DNA, respectively.

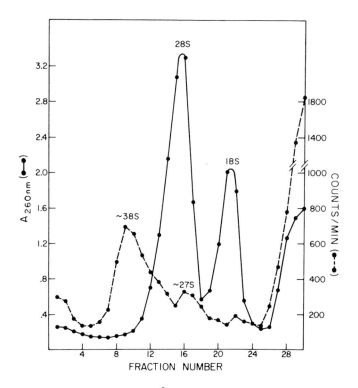

Fig. 6 - Incorporation of ^{3}H-thymidine into nucleic acid by artificially activated anucleate fragments.

Artificially activated anucleate fragments were incubated for 1 hour in 30 ml of artificial sea water at 18°C with 10 µCi/ml of ^{3}H-thymidine. The fragments were washed and phenol extracted, as specified in Materials and Methods. A portion of the dried nucleic acid was suspended in 0.005 M EDTA, 0.275 M KCl and 0.05 M Tris, pH 7.5. 0.4 ml of the solution was layered onto 11 ml of a 5 - 20 % (wt/vol) sucrose gradient plus a 1 ml 60 % sucrose pad all in the same buffer as the nucleic acid. The gradient was centrifuged at 40,000 rpm for 4 hours at 2°C in a S.W. 41 rotor. 0.4 ml fractions were collected. Ultra violet light absorption at 260 nm was read and 0.3 ml fractions were dried on filter papers and washed with ice cold TCA, as specified in Methods.

$A_{260\ nm}$ ●——●   Radioactivity ●-----●

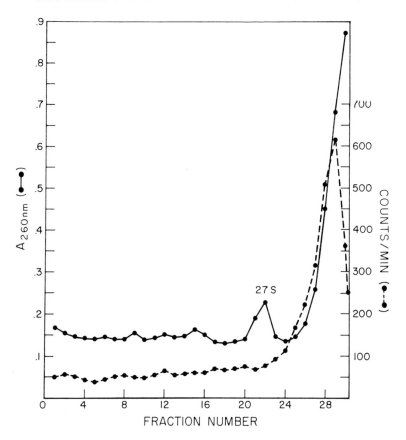

Fig. 7 - The sucrose sedimentation of the lysate of isolated mitochondria from ^{3}H-thymidine labeled anucleate fragments.

Artificially activated anucleate fragments were incubated for 2.5 hours in a total volume of 40 ml with 12 µCi ^{3}H-thymidine/ml. Mitochondria were isolated and prepared for sedimentation analysis, as specified in Materials and Methods. The mitochondria lysate was layered onto 12 ml of a 5 - 20 % (wt/wt) sucrose gradient in 0.275 M KCl, 0.01 M MgAc and 0.01 M Tris (pH 7.5) and centrifuged at 40K for 4 hours at 2°C in a S.W. 41 rotor. The 27 S O.D. peak is assumed to be that of the nicked mitochondrial DNA.

Radioactivity ●-----● A_{260} nm   ●———●

SYNTHESIS OF CLOSED CIRCULAR DNA IN FERTILIZED AND ACTIVATED EGGS AND IN ANUCLEATE FRAGMENTS OF THE SEA URCHIN ARBACIA LIXULA

Horst Bresch

Stazione Zoologica, Naples, Italy,
and
Medizinische Hochschule Hannover,
Abteilung für experimentelle
Pathologie, Hannover, Germany

The finding that mitochondria of higher animals contain circular DNA which replicates in vivo and in vitro (general references : Roodyn and Wilkie, 1968; Nass, 1969; Ashwell and Work, 1970) raises the question to what extent the turnover of these macromolecules is determined by the nucleus. To investigate this problem systems may be used in which the direct influence of the nucleus on replication has been cut off. An anucleate cell fragment, still containing mitochondria and still being able to live for a certain time, can serve as a model. Thus the turnover of mitochondrial DNA in the anucleate fragment may be compared with the turnover in whole cells kept under the same conditions.

For biochemical studies larger amounts of cells, which lend themselves to anucleation are required. Eggs of the sea urchin Arbacia lixula have been found to be a suitable source for such experiments. As anucleate fragments cleave after fertilization their biological activity can easily be tested. It has also been demonstrated by Geuskens (1965) that the anucleate fragment of the Arbacia lixula egg contains all visible ultrastructural constituents as the whole egg. Bibring, Brachet, Gaeta and Graziosi (1965) isolated a DNA from these fragments which was characterized by its buoyant density. In experiments done with non-activated as well as activated anucleate fragments of this species, Baltus, Quertier, Ficq and Brachet (1965) observed incorporation of thymidine into an acid precipitable material.

The experiments reported here were designed to answer the following questions :

1. Does replication of closed circular DNA occur in fertilized and in activated eggs ?
2. If this is the case what is the rate of its turnover in the corresponding anucleate fragments ? What part does the nucleus play with regard to the replication of circular DNA ?

Materials and Methods

Collection, anucleation and incubation of eggs

The animals were collected from the bay of Naples. Eggs and sperm were obtained by injection of 0.5 ml of 0.5 M-KCl into the animals. The eggs were washed three times in 0.2 µ Millipore-filtered sea water to which 100 U penicillin G and 0.1 mg streptomycine sulfate/ml had been added (SSP). All experiments were performed at 18°C. Separation of eggs into white nucleate and red anucleate parts was achieved by centrifugation at 11000 rpm for 2-4 min. in the SW 25-rotor of the Spinco model L 50. The tube contained : 9 ml 1.1 M-sucrose pH 7.8 as subphase, 15 ml 0.75 M-sucrose made by dilution of 1.1 M-sucrose with sea water as medium phase and 5 ml sea water in which intact unfertilized eggs had been suspended as top phase (Harvey, 1956; Baltus et al., 1965). After the centrifugation, the red halves floated on the subphase and the white halves between the top and middle phases. Only those gradients were used in which the middle phase was free of whole eggs or broken parts. After centrifugation, the split eggs were washed twice with SSP, examined under a stereo microscope, and counted in capillary tubes. The suspension of anucleate parts was not contaminated with nucleate fragments when checked under the microscope. For each experiment , total and split eggs of different animals were pooled and afterwards divided into three equal aliquots. One aliquot was fertilized, another one activated, and the third one was left untreated. Two minutes after fertilization the cells were sedimented in a hand centrifuge, washed with SSP and incubated in the same medium. Activation was achieved by treatment with a solution of 3% NaCl in SSP for 15 min. The washed cells were incubated in the same way as the untreated and fertilized eggs.

Incorporation of thymidine into the soluble pool and into total DNA

10^5 untreated, activated or fertilized eggs were suspended in 5 ml SSP + 1 μc thymidine-methyl-H^3 (New England Nuclear Corp., 16c/mM) and incubated in a 25 ml test tube which was tilted rhythmically on a tray. After 40, 80, 120 and 160 min. the cells of one test tube were sedimented. The sediment was immediately resuspended in ice cold 0.5 M-NaCl-KCl solution 19:1, rapidly followed by centrifugation in a hand centrifuge. The sediment was homogenized in 10 ml 0.2 N-H_2SO_4 at 0°C. The homogenate was centrifuged at 30000 xg for 20 min. The supernatant was then sucked through a 0.45 μ Millipore filter and an aliquot of 0.05 ml was taken and measured in a toluene-methanol scintillator (4 ml methanol + 10 ml toluene scintillator, 4 g PPO + 0.1 g dimethyl POPOP/1 1 toluol) in a Packard liquid scintillation spectrometer. The sediment was washed with ethanol, dried and burned in an O_2 atmosphere according to the method of Kalberer and Rutschmann (1961). The tritiated water was also measured in the toluene-methanol scintillator.

Incorporation of thymidine into mitochondrial DNA

For each experiment, about 10^6 eggs or fragments were incubated in 100 ml SSP containing 2 μc thymidine-methyl-H^3/ml. The suspension was slowly agitated with a plastic stirrer. After incubation the suspended eggs were washed with NaCl-KCl solution, resuspended in a mixture of 0.3 M-sucrose 0.36 M-KCl, 0.03 M-Tris-HCl and 0.003 M-EDTA pH 7.6 (Piko, Tyler and Vinograd, 1967) at 0°C, and homogenized at 500 rpm in an ice cold Potter-Elvehjem homogenizer with a teflon pistil, with the efficiency being periodically checked under the microscope. The homogenate was centrifuged at 800 xg for 15 min. Mitochondria and yolk platelets, together with small amounts of pigment, were sedimented at 9000 xg. The sediment was lysed in 0.5 ml of 2% SDS, 0.01 M-Tris-HCl, 0.001 M-EDTA at pH 7.5 and spun at 43000 xg in a CsCl-ethidium bromide gradient according to Radloff, Bauer and Vinograd (1967).

The ethidium bromide DNA complex was visualized in the near UV.

For fractionation the gradient was pumped upwards in a special apparatus which will be described elsewhere and two drops were collected in one fraction. Fluorescence of the fractions was measured at 590 nm in a Zeiss spectrophotometer PMQ II with the fluorescent attachment ZFM 4 (excitation light 546 nm), 100 µg bovine serum albumin were added and the DNA was precipitated in the cold with 20% trichloracetic acid (TCA). The precipitate was collected on 0.45 µ Millipore filter, washed with 5% TCA, dried and counted in Omnifluor-toluene scintillator (New England Nuclear Corp., 4 g Omnifluor /1 1 toluene.)

When dye free DNA was investigated, the dye was removed with Dowex 50 WX 4, 200-400 mesh, Na-form.

DNase treatment of the mitochondria-yolk platelet sediment

The sediment was suspended in 1 ml of a solution consisting of 0.5 M-sucrose; 0.005 M-$MgCl_2$; 0.01 M-TRIS-HCl pH 7.4 and 50 µgDNase (Desoxyribonuclease 1, Serva Heidelberg) and incubated for 30 min. at 25°C. The reaction was stopped in ice and 9 ml of 0.3 M-sucrose; 0.36 M-KCl; 0.03 M-EDTA; 0.01 M-TRIS-HCl pH 7.4 were added and the particles sedimented at 9000 xg. The sediment was washed again in this solution and after another centrifugation lysed as described above.

Equilibrium density banding

To 1 ml 0.01 M-TRIS-HCl pH 8.0 containing about 0.5 µg DNA, 0.8 µg of Micrococcus lysodeicticus DNA was added as a reference (ρ : 1.731 g/ml); the solution was adjusted to a density of 1.70 g/cm^3 with CsCl and spun at 44770 rpm for 20 hours in a standard 12 mm Kel-F cell of the rotor AN-D in a Spinco model E. The photographs taken at 260 nm were scanned with a Joyce Loebl microdensitometer.

Electron microscopy

The mitochondria-yolk platelet pellet was fixed at 0°C in a solution of 2% osmium tetroxide in 0.1 M 2,4,6-collidin-buffer (Merck) + 40% sucrose pH 7.4, 1300 m osmols, for 4 hours, dehydrated in ethanol, embedded over propylene-oxide in Durcupan (Fluka) and cut with a Reichert ultramicrotome, diamond knife Om Cl_2. The ultra-thin sections were stained in lead citrate (Venable and Coggeshall, 1965).

0.05 ml of the CsCl-ethidium bromide-DNA fractions were diluted with a tenfold volume of 1 M-ammonium acetate; 100 µg cytochrome c (Boehringer) were added and the solution spread over a hypophase of 0.25 M-ammonium acetate (Kleinschmidt, 1968). The film was transferred to carbon-coated formvar-copper grids, dried and shadowed with platinum while rotating under an angle of 6°. The grids were examined in a Philips EM 200 electron microscope. The circumference of photographed molecules was measured with a perlon thread which was fixed to the illustrated molecules with glue. No shrinkage of the thread was observed after fixation.

Results and discussion

Biological observations

After fertilization most of whole eggs developed up to the 4-cell stage within 160 minutes, about 30% of the embryos contained already eight blastomeres. Activation for 15 minutes caused raising of the fertilization membrane in more than 90% of the eggs. After 160 minutes, 50% of the embryos had developed up to the 4-cell stage and 20% had cleaved only once. Under these conditions the nucleate fragments behaved in the same way as whole eggs. In fertilized anucleate halves the development was slower. 70% of the fragments had still only divided once 160 minutes after fertilization. In contrast, activated anucleate fragments cleaved only rarely; 160 minutes after activation more than 90% of the cells were still uncleaved and most of the cleaved fragments contained only two unequal blastomeres.

Thymidine-incorporation studies

Labelled thymidine was used to detect a synthesis of total DNA at various times after fertilization or activation. The incubated eggs or fragments were homogenized in 0.2 N-sulfuric acid. The homogenate was spun down and the radioactivity of supernatant and precipitate measured. The radioactivity of the supernatant corresponds to the amount of thymidine, or a metabolite of it, present in the soluble pool. The sediment obtained at 30000 xg contains the radioactivity of the nucleoside which had been incorporated into DNA. As Nemer (1962) demonstrated, unfertilized eggs of Paracentrotus lividus take up only small amounts of thymidine, with the amount of incorporation increasing after fertilization. Unfertilized eggs of Arbacia lixula also take up some thymidine, a plateau being reached after 40 minutes. Most of the nucleoside is found in the soluble pool. In the sediment only a very low amount of radioactivity is detected, indicating very little, if any, DNA synthesis (Fig.1a). After fertilization, the uptake of thymidine by the egg increases; in the soluble pool a plateau is reached after 40 minutes, whereas the radioactivity in the acid insoluble sediment increases steadily (Fig.1a). Activated whole eggs also synthesize DNA (Fig.1a). Compared to fertilized eggs, the lower extent may be explained by the slower developing rate. If left untreated, nucleate halves also take up small amounts of the precursor (Fig. 1b). As in the whole eggs, the radioactivity in the acid insoluble fraction is very low. If activated, the influx of thymidine into nucleate fragments increases as in whole eggs; DNA is also synthesized (Fig.1b). Untreated anucleate fragments show the same pattern as the white parts (Fig. 1b). Activated anucleate fragments also take up more thymidine, but in contrast to white halves it is not incorporated into DNA (Fig.1b). Apparently no measurable DNA synthesis occurs in anucleate activated eggs.

Taking into account a possible low turnover of circular DNA in the egg and anucleate half after activation or fertilization, in further experiments a tenfold more eggs and halves were used than before, and attempts were made to isolate circular DNA in a CsCl-ethidium bromide gradient (Radloff et al., 1967) and to determine the incorporation of thymidine into the macromolecules. Eggs or fragments were incubated in the same medium as before and the development stopped at certain stages. The cells were homogenized in a sucrose-KCl-TRIS buffer and submitted to differential centrifugation. The 800 xg sediment contained mainly unbroken cells, membranes, and pigment granules. Mitochondria together with yolk platelets were sedimented at 9000 xg. A pellet obtained from anucleate fragments was examined in the electron microscope and consisted mainly of yolk platelets and mitochondria laying between (Fig.2). Further separation of mitochondria from yolk platelets was tried in a sucrose gradient. The separation, however, was not quantitative and a part of the mitochondria was lost in the yolk fraction. In order not to lose a considerable portion of mitochondria in experiments done

ANUCLEATE SYSTEMS: BACTERIA AND ANIMAL CELLS

with anucleate fragments attemps of further separation were omitted. The sediment was lysed and the DNA isolated in a CsCl-ethidium bromide gradient. This technique permits separation of linear from closed circular DNA. The ethidium bromide-DNA intercalation complexes of linear molecules move into the less dense band as do broken circular molecules, whereas closed circular molecules are banded together at a slightly higher density in the presence of the dye. This difference is sufficient for separate collection of both bands. A third band, which is sometimes observed between the light and heavy band, consists of so called catenated molecules (Hudson and Vinograd, 1967); two or three circular molecules are interlocked like links in a chain. This middle band was observed in gradients of whole eggs but because of the smaller quantity of lysate used less often in gradients of anucleate fragments. In all gradients the band consisting of closed circular molecules was present. In general all three bands of the gradients of whole eggs and embryos contained a total of between one and four µg of DNA. Thirty to sixty percent of this amount was present in the lower band as calculated from the UV absorbance of dye free DNA. The weak middle band amounted to 5 to 10%. The gradients of anucleate halves contained only about one third of the quantity present in gradients of whole eggs. These values certainly cannot be taken to represent the eggs' total content of circular DNA since during the homogenisation and fractionation procedures DNA containing particles may have been lost. For characterization of the DNA in the heavy bands, two samples were taken, one from whole eggs, the other from anucleate halves, and prepared for electron microscopy. More than 90% of the examined molecules consisted of partially twisted or relaxed circles, the latter probably having been broken during the preparation procedure. The mean length distribution corresponds to a circumference of 4.6 µ (Fig.3).

The heavy band DNA from two other gradients of whole eggs and anucleate halves were freed from ethidium bromide and spun to equilibration in a CsCl equilibrium density gradient. According to the photographs, both samples contained only one kind of DNA banding at a density of 1.700 g/cm^3 (Fig.4). This is in accordance with the findings of Bibring et al. (1965).

It is still not certain if yolk platelets of the sea urchin egg contain DNA. Piko et al. (1967) claimed the occurrence of DNA in these particles assuming the same amount as in mitochondria. In a later paper (Piko, Blair, Tyler and

Vinograd, 1968), however, the localization of 80-90% of cytoplasmic DNA in mitochondria was proposed. In any case, the heavy band in the CsCl-ethidium bromide gradient contains mitochondrial DNA. If yolk platelet DNA is present in this band, it must have at least similar physical properties to mitochondrial DNA, since the band contains only closed circular molecules with about the same circumference and since only a single band was present in the analytical CsCl gradient.

To investigate the turnover of circular DNA the eggs were incubated in thymidine-H^3 after having been fertilized. Later, when different developmental stages were reached, a sample was taken and the DNA was isolated from the mitochondria-yolk platelet sediment as described above. Figure 5 illustrates the localization of linear and circular DNA in the CsCl-ethidium bromide gradients. The fluorescence and the TCA insoluble radioactivity of the fractions are indicated. Fig.5 a,b,c, show the increase of the radioactivity in the different DNA fractions from fertilized eggs measured until the 32-cell stage was reached. In all stages all three DNA bands are labelled. The high radioactivity in the bands containing theoretically linear as well as broken circular DNA (light band) and the comparativeley low labelling in the bands containing catenated molecules and closed circular monomers (heavy bands) is striking. The great difference in specific activities between the DNA in the light and heavy bands indicates that the light bands must contain besides broken circular molecules from the same origin as those of the heavy bands, at least one other kind of DNA which turns over to a multifold higher rate than the circular DNA. There are two explanations for this fact :

1. The linear DNA originates from yolk platelets.
2. The band still contains a linear DNA, most likely nuclear DNA which stuck to mitochondria or yolk platelets after the eggs had been homogenized.

As already mentioned the question as to whether or not yolk platelets of the sea urchin egg contain DNA which replicates after fertilization is not yet settled. Anderson (1969) studying nuclear and cytoplasmic DNA synthesis by high resolution autoradiography detected an incorporation of thymidine into mitochondria and yolk platelets of fertilized Paracentrotus lividus eggs, the majority of the silver grains in the cytoplasm being located over mitochondria. The labelling is not washed out by simple rinsing but disappears after

DNase treatment. In addition to a replication of mitochondrial DNA also a replication of some DNA in the yolk is discussed. Working with the polychaete Ophryotrocha labronica Emanuelsson (1971) reported interesting results as to origin and turnover of yolk platelet DNA in this species. He demonstrated with autoradiographic methods and with liquid scintillation measurements that most of yolk DNA consists of nuclear DNA which, during vitellogenesis is expelled from the oocyte and incorporated in the yolk granules. He further showed that during early development this DNA is broken down and diminishes rapidly.

To see whether a DNA sticks firmly to the particles, the mitochondria-yolk platelet sediment isolated from 4-8-cell stage embryos was treated with DNase, lysed and spun in a CsCl-ethidium bromide gradient. Fig. 5d demonstrates that most of the radioactivity in the light band disappeared, whereas the radioactivity of both circular types was unchanged. If high labelled linear yolk platelet DNA was present it was broken by the DNase treatment. Recently the author isolated a mitochondria fraction essentially free of yolk platelets from early stages of Psammechinus miliaris embryos and found also a high radioactivity in the light band when the embryos had been incubated in thymidine-H^3 before.
After DNase treatment of the mitochondria fraction most of the radioactivity in the light band also disappeared. According to this experiment the highly labelled linear DNA might originate from another source than yolk, though it cannot be excluded that during homogenisation DNA from broken yolk platelets stuck to mitochondria. Since it is known that nuclear DNA is rapidly synthesized after fertilization it is conceivable that this DNA partially set free during homogenisation from broken nuclei and sticking firmly to mitochondria and perhaps also to yolk platelets causes the high radioactivity in the light bands.

As to the thymidine incorporation in both circular DNA molecules, it is remarkable that the catenated molecules are labelled to a higher degree than the monomers. If interlocked dimers are formed by a double recombination process between two circular monomers as proposed by Hudson and Vinograd (1967) it is scarcely conceivable that the specific activity of these molecules should exceed that of the parental monomers. Considering the comparatively low labelling of both circular DNA fractions the assumption arises that the circular molecules were not labelled in the course of a true replication step but rather by a repair mechanism, the catenated molecules being repaired to a higher degree. Since it is unknown, however, whether or not only a small fraction of total mitochondrial DNA may be replicated in the egg it is not possible to make one among different hypotheses more probable.

If the eggs were not fertilized but activated in hypertonic sea water and thereafter continuously incubated in thymidine-H^3 as fertilized eggs were, the nucleoside also was incorporated in all three DNA fractions. Fig. 5e illustrates the profile of the gradient of parthenogenetic activated eggs, about 30% had developed up to the 4-cell stage. The specific radioactivities of all DNA fractions are lower than those of gradients from fertilized eggs which had been incubated for the same time. The slower developing rate may explain this fact. If incubated for the same time, fertilized anucleate fragments also incorporate thymidine into all three DNA fractions (Fig. 5f). Contrary to fertilized anucleate halves, the nucleoside was not, or only to a very small extent, incorporated into the circular DNA of activated anucleate fragments, as illustrated by Fig. 5 g,h,i, thus confirming the results obtained by precipitation of "total" DNA. A very small amount of radioactivity was detected in fractions of circular monomers 200 minutes after continuous incubation in thymidine-H^3. After that time about 10% of the fragments had divided once or twice, but almost all of the cleaved cells were abnormal. The low labelling of circular monomers (catenated molecules were not observed in this gradient) might originate from either cleaved fragments completely lacking nuclear DNA or from fragments still containing a part of the nuclear genome. The labelling found in the light bands of all three gradients might be caused by replication or repair of these residual nuclear DNA molecules after activation provided the labelling does not originate from possibly existing metabolically active yolk DNA.

According to the finding that thymidine is not or only scarcely incorporated into mitochondrial DNA (M-DNA) of activated anucleate fragments the incorporation of the nucleoside seems to be under nuclear control. Regulation by the nucleus may be exerted via transcription of an RNA and finally via a protein determining the turnover or repair of the DNA in mitochondria. It is also possible, however, that the nucleus determines the replication of M-DNA in a more indirect way by influencing the division rate of the cell

which itself might in some way be linked to the control of
M-DNA metabolism. The lower rate of thymidine incorporation
into the circular DNA of activated eggs and fertilized anucleate fragments compared to fertilized eggs as well as the very
small labelling of these molecules in anucleate fragments
might be explained as a consequence of a lower division rate.
The question whether or not a cell without a nucleus is able
to divide regularly or not is not yet settled. Harvey (1956)
described anucleate embryos of Arbacia punctulata which,
according to her observations, contained up to 500 cells.
Unfortunately, in most further publications dealing with the
anucleate sea urchin eggs, no comment is given about the degree of cleavage obtained. The degree to which an M-DNA synthesis occurs in non-dividing cells has so far been poorly
investigated. Berger and Schweiger (1969) measured a synthesis of M-DNA in anucleate cells of Acetabularia mediterranea.
Acetabularia itself is a peculiar cell able to regenerate
and even differentiate in the absence of the nucleus. Results
obtained in experiments on this plant cell are only cautious-
ly transferrable to other systems. Mezger-Freed (1963) detected incorporation of labelled thymidine into acid insoluble
material in anucleate eggs of Rana pipiens. Whether the nucleoside is localized in the M-DNA or not has not been demonstrated yet. To get a better insight into the regulation of
M-DNA synthesis in non-dividing long living animal cells,
nerve cells may prove to be a suitable object to work with.
M-DNA synthesis in a giant axon is under current investigation in this laboratory.

Summary

Total DNA of eggs and nucleate halves of the sea urchin
Arbacia lixula is already replicated 40 minutes after fertilization or parthenogenetic activation as demonstrated by
incorporation of thymidine-methyl-H^3 into acid precipitable
material. The nucleoside, however, was not detected in the
precipitate from activated anucleate fragments. Mitochondria
together with yolk platelets were isolated from fertilized
and activated eggs and from anucleate fragments after incubation in thymidine. In a CsCl-ethidium bromide gradient
closed circular DNA was isolated from the mitochondria-yolk
platelet sediment. A labelling of this DNA was observed in
fertilized and activated eggs and in fertilized anucleate
fragments. In activated anucleate fragments, however, the

DNA is not significantly labeled . According to these results the anabolism of circular DNA in the egg of <u>Arbacia lixula</u> is under nuclear control.

Acknowledgement

I wish to express my gratitude to Prof. J.Brachet who introduced me into the field of embryology and with whom I had the opportunity to discuss this work. I also thank Dr. Adelheid Emminger, Medizinische Hochschule Hannover, who helped me in preparing anucleate fragments, Dr. Rainer Martin Stazione Zoologica Naples, for electron microscopy and Jacques Verhulst, Université Libre de Bruxelles, Département de Biologie Moléculaire, Rhode-St-Genèse, for running analytical CsCl gradients.

This work was kindly supported by Euratom and by the Deutsche Forschungsgemeinschaft.

References

Anderson,A.W. (1969) - J.Ultrastruct.Res. 26, 95.
Ashwell,M. and Work,T.S. (1970) - Annual Review of Biochemistry 39, 251.
Baltus,E., Quertier,J., Ficq,A. and Brachet,J. (1965) - Biochim. biophys. Acta 95, 408.
Berger,S. and Schweiger,H.G. (1969) - Physiol.Chem. and Physics 1, 280.
Bibring,T., Brachet,J., Gaeta,S.F. and Graziosi,F. (1965) - Biochim. biophys. Acta 108, 644
Emanuelsson,H. (1971) - Z. Zellforsch. 113, 450.
Geuskens,M. (1965) - Exp. Cell Res. 39, 413.
Harvey,E.B. (1956) - The American Arbacia and other Sea Urchins. Princeton : Princeton University Press, Princeton, New Jersey.
Hudson,B. and Vinograd,J. (1967) - Nature (London) 216, 647.
Kalberer,F. and Rutschmann,J. (1961) - Helv. chim. Acta 44, 1956.
Kleinschmidt,A.K. (1968) - In : Methods in Enzymology ed. by L. Grossman and K. Moldave, vol. XII Part B New York and London, Academic Press.
Mezger-Freed,L. (1963) - J. Cell Biol. 18, 471.
Nass,M.M.K. (1969) - Science 165, 25.
Nemer,M. (1962) - J. Biol. Chem. 237, 143.
Piko,L., Tyler,A. and Vinograd,J. (1967) - Biol. Bull. 132, 68.

Piko,L., Blair,G.D., Tyler,A. and Vinograd,J. (1968) - Proc. Nat. Acad. Sci. Wash. $\underline{59}$, 838.
Radloff,R., Bauer,W. and Vinograd,J. (1967) - Proc. Nat. Acad. Sci. Wash. 57, 1514.
Roodyn,D.B. and Wilkie,D. (1968) The Biogenesis of Mitochondria. London : Methuen & Co. Ltd., London.
Venable,J.H. and Coggeshall,R.A. (1965) - J. Cell Biol. $\underline{25}$, 407.

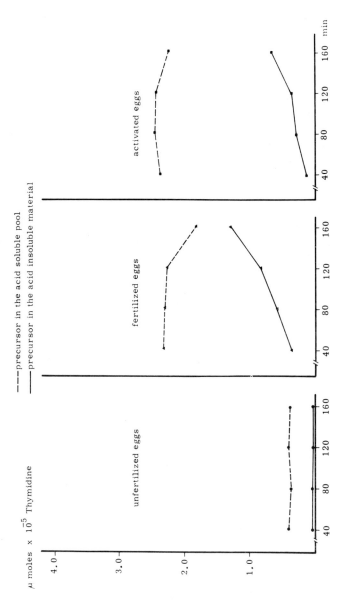

Fig. 1a Incorporation of thymidine-methyl-H^3 into *Arbacia lixula* eggs and early embryos. Continuous incubation was performed and samples taken at indicated times. The values represent the uptake of the nucleoside by 10^5 eggs.

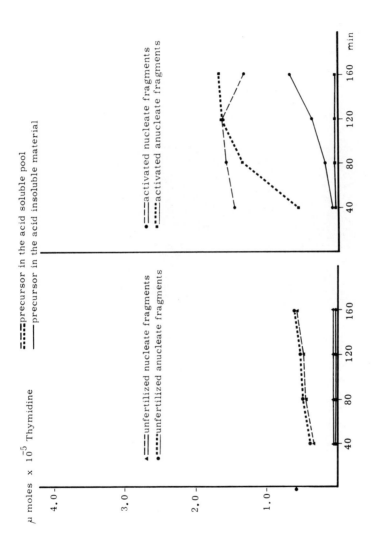

Fig. 1b Incorporation of thymidine-methyl-H^3 into nucleate and anucleate fragments of Arbacia lixula. Halves were incubated as whole eggs. The values represent the uptake of the nucleoside by 10^5 fragments.

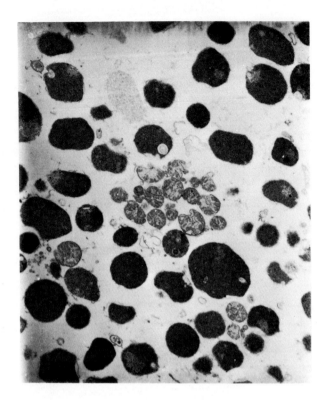

Fig. 2 Mitochondria-yolk platelet sediment of anucleate fragments from Arbacia lixula eggs. The sediment was obtained by differential centrifugation at 9000 xg. x 11200.

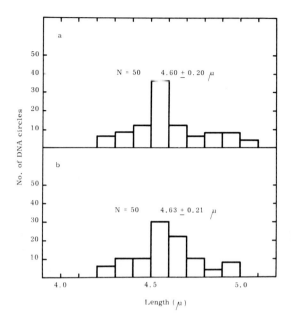

Fig. 3   Length distribution of circular DNA molecules isolated by CsCl-ethidium bromide density gradient centrifugation.
   a. DNA from whole eggs
   b. DNA from anucleate fragments of <u>Arbacia lixula</u> eggs.
The standard deviation is indicated.

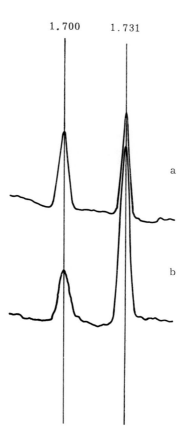

Fig. 4 CsCl density centrifugation of circular DNA from eggs and fragments of <u>Arbacia lixula.</u> DNA of heavy bands in CsCl-ethidium bromide gradients was freed from ethidium bromide and centrifuged for 20 hr at 44770 rpm/min. Upper graph : DNA from whole eggs, lower graph : DNA from anucleate fragments. The band to the right is <u>Micrococcus lysodeicticus</u> DNA with a density of 1.731 g/cm^3.

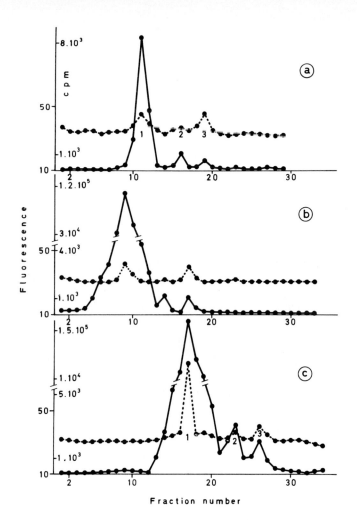

Fig.5a-i CsCl-ethidium bromide density gradients of mitochondria -yolk platelet lysates from <u>Arbacia lixula.</u> Only those fractions are indicated which are close to the bands. The top of the tube is to the left, the bottom to the right in the graph. From left to right : light band containing linear DNA and open circles (1). The next two bands contain catenated molecules (2) and closed circular monomers (3). Interrupted line : Fluorescence in arbitrary units. Full line : Radioactivity.

    a-c fertilized eggs, after continuous incubation in thymidine-methyl-H^3. a 2-cell stage, 90 min. b 4- and 8-cell stage, 140 min. c mainly 32-cell stage, 200 min.

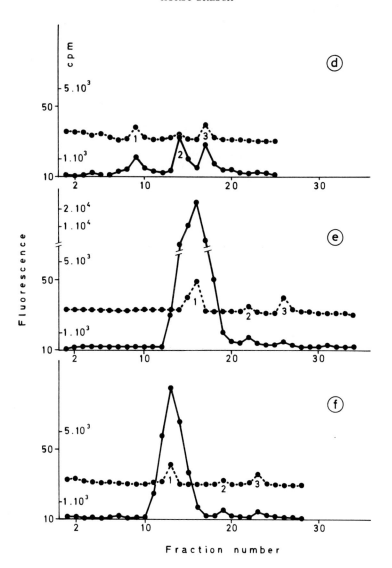

d–f 140 minutes after continuous incubation in thymidine-methyl-H^3. d fertilized eggs 4- and 8-cell stage. Gradient of a DNase treated mitochondria-yolk platelet sediment. e activated eggs, 30% in the 4-cell stage. f fertilized anucleate fragments, 50% in the 2-cell stage.

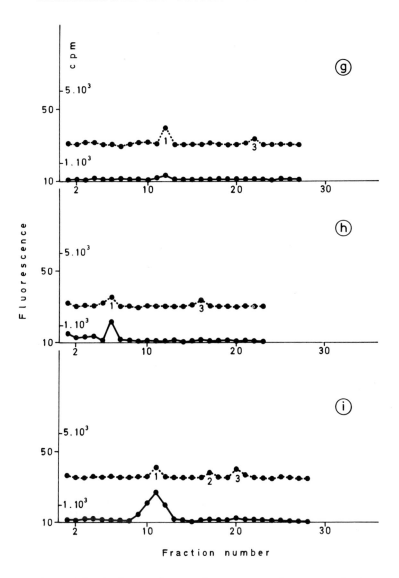

g–i activated anucleate fragments, after continuous incubation in thymidine-methyl-H^3. g 90 min. h 140 min. i 200 min.